Topics in Current Chemistry

Editorial Board:
K.N. Houk • C.A. Hunter • M.J. Krische • J.-M. Lehn
S.V. Ley • M. Olivucci • J. Thiem • M. Venturi • P. Vogel
C.-H. Wong • H. Wong • H. Yamamoto

Topics in Current Chemistry
Recently Published and Forthcoming Volumes

Novel Sampling Approaches in Higher Dimensional NMR
Volume Editors: Martin Billeter, Vladislav Orekhov
Vol. 316, 2012

Advanced X-Ray Crystallography
Volume Editor: Kari Rissanen
Vol. 315, 2012

Pyrethroids: From Chrysanthemum to Modern Industrial Insecticide
Volume Editors: Noritada Matsuo, Tatsuya Mori
Vol. 314, 2012

Unimolecular and Supramolecular Electronics II
Volume Editor: Robert M. Metzger
Vol. 313, 2012

Unimolecular and Supramolecular Electronics I
Volume Editor: Robert M. Metzger
Vol. 312, 2012

Bismuth-Mediated Organic Reactions
Volume Editor: Thierry Ollevier
Vol. 311, 2012

Peptide-Based Materials
Volume Editor: Timothy Deming
Vol. 310, 2012

Alkaloid Synthesis
Volume Editor: Hans-Joachim Knölker
Vol. 309, 2012

Fluorous Chemistry
Volume Editor: István T. Horváth
Vol. 308, 2012

Multiscale Molecular Methods in Applied Chemistry
Volume Editors: Barbara Kirchner, Jadran Vrabec
Vol. 307, 2012

Solid State NMR
Volume Editor: Jerry C. C. Chan
Vol. 306, 2012

Prion Proteins
Volume Editor: Jörg Tatzelt
Vol. 305, 2011

Microfluidics: Technologies and Applications
Volume Editor: Bingcheng Lin
Vol. 304, 2011

Photocatalysis
Volume Editor: Carlo Alberto Bignozzi
Vol. 303, 2011

Computational Mechanisms of Au and Pt Catalyzed Reactions
Volume Editors: Elena Soriano, José Marco-Contelles
Vol. 302, 2011

Reactivity Tuning in Oligosaccharide Assembly
Volume Editors: Bert Fraser-Reid, J. Cristóbal López
Vol. 301, 2011

Luminescence Applied in Sensor Science
Volume Editors: Luca Prodi, Marco Montalti, Nelsi Zaccheroni
Vol. 300, 2011

Chemistry of Opioids
Volume Editor: Hiroshi Nagase
Vol. 299, 2011

Electronic and Magnetic Properties of Chiral Molecules and Supramolecular Architectures
Volume Editors: Ron Naaman, David N. Beratan, David H. Waldeck
Vol. 298, 2011

Natural Products via Enzymatic Reactions
Volume Editor: Jörn Piel
Vol. 297, 2010

Nucleic Acid Transfection
Volume Editors: Wolfgang Bielke, Christoph Erbacher
Vol. 296, 2010

Novel Sampling Approaches in Higher Dimensional NMR

Volume Editors: Martin Billeter · Vladislav Orekhov

With Contributions by

H. Arthanari · R. Freeman · S. Hiller · J.C. Hoch · S.G. Hyberts ·
K. Kazimierczuk · W. Koźmiński · E. Kupče · M.W. Maciejewski ·
M. Misiak · M. Mobli · A.D. Schuyler · J. Stanek · A.S. Stern ·
G. Wagner · G. Wider · A. Zawadzka-Kazimierczuk

Editors
Prof. Dr. Martin Billeter
Department of Chemistry &
Molecular Biology
University of Gothenburg
Göteborg
Sweden

Prof. Dr. Vladislav Orekhov
Swedish NMR Centre
University of Gothenburg
Göteborg
Sweden

ISSN 0340-1022 e-ISSN 1436-5049
ISBN 978-3-642-27159-5 e-ISBN 978-3-642-27160-1
DOI 10.1007/978-3-642-27160-1
Springer Heidelberg Dordrecht London New York

Library of Congress Control Number: 2012932900

© Springer-Verlag Berlin Heidelberg 2012

This work is subject to copyright. All rights are reserved, whether the whole or part of the material is concerned, specifically the rights of translation, reprinting, reuse of illustrations, recitation, broadcasting, reproduction on microfilm or in any other way, and storage in data banks. Duplication of this publication or parts thereof is permitted only under the provisions of the German Copyright Law of September 9, 1965, in its current version, and permission for use must always be obtained from Springer. Violations are liable to prosecution under the German Copyright Law.

The use of general descriptive names, registered names, trademarks, etc. in this publication does not imply, even in the absence of a specific statement, that such names are exempt from the relevant protective laws and regulations and therefore free for general use.

Printed on acid-free paper

Springer is part of Springer Science+Business Media (www.springer.com)

Volume Editors

Prof. Dr. Martin Billeter
Department of Chemistry &
Molecular Biology
University of Gothenburg
Göteborg
Sweden

Prof. Dr. Vladislav Orekhov
Swedish NMR Centre
University of Gothenburg
Göteborg
Sweden

Editorial Board

Prof. Dr. Kendall N. Houk

University of California
Department of Chemistry and Biochemistry
405 Hilgard Avenue
Los Angeles, CA 90024-1589, USA
houk@chem.ucla.edu

Prof. Dr. Christopher A. Hunter

Department of Chemistry
University of Sheffield
Sheffield S3 7HF, United Kingdom
c.hunter@sheffield.ac.uk

Prof. Michael J. Krische

University of Texas at Austin
Chemistry & Biochemistry Department
1 University Station A5300
Austin TX, 78712-0165, USA
mkrische@mail.utexas.edu

Prof. Dr. Jean-Marie Lehn

ISIS
8, allée Gaspard Monge
BP 70028
67083 Strasbourg Cedex, France
lehn@isis.u-strasbg.fr

Prof. Dr. Steven V. Ley

University Chemical Laboratory
Lensfield Road
Cambridge CB2 1EW
Great Britain
Svl1000@cus.cam.ac.uk

Prof. Dr. Massimo Olivucci

Università di Siena
Dipartimento di Chimica
Via A De Gasperi 2
53100 Siena, Italy
olivucci@unisi.it

Prof. Dr. Joachim Thiem

Institut für Organische Chemie
Universität Hamburg
Martin-Luther-King-Platz 6
20146 Hamburg, Germany
thiem@chemie.uni-hamburg.de

Prof. Dr. Margherita Venturi

Dipartimento di Chimica
Università di Bologna
via Selmi 2
40126 Bologna, Italy
margherita.venturi@unibo.it

Prof. Dr. Pierre Vogel

Laboratory of Glycochemistry
and Asymmetric Synthesis
EPFL – Ecole polytechnique féderale
de Lausanne
EPFL SB ISIC LGSA
BCH 5307 (Bat.BCH)
1015 Lausanne, Switzerland
pierre.vogel@epfl.ch

Prof. Dr. Chi-Huey Wong

Professor of Chemistry, Scripps Research Institute
President of Academia Sinica
Academia Sinica
128 Academia Road
Section 2, Nankang
Taipei 115
Taiwan
chwong@gate.sinica.edu.tw

Prof. Dr. Henry Wong

The Chinese University of Hong Kong
University Science Centre
Department of Chemistry
Shatin, New Territories
hncwong@cuhk.edu.hk

Prof. Dr. Hisashi Yamamoto

Arthur Holly Compton Distinguished Professor
Department of Chemistry
The University of Chicago
5735 South Ellis Avenue
Chicago, IL 60637
773-702-5059
USA
yamamoto@uchicago.edu

Topics in Current Chemistry
Also Available Electronically

Topics in Current Chemistry is included in Springer's eBook package *Chemistry and Materials Science*. If a library does not opt for the whole package the book series may be bought on a subscription basis. Also, all back volumes are available electronically.

For all customers with a print standing order we offer free access to the electronic volumes of the series published in the current year.

If you do not have access, you can still view the table of contents of each volume and the abstract of each article by going to the SpringerLink homepage, clicking on "Chemistry and Materials Science," under Subject Collection, then "Book Series," under Content Type and finally by selecting *Topics in Current Chemistry*.

You will find information about the

– Editorial Board
– Aims and Scope
– Instructions for Authors
– Sample Contribution

at springer.com using the search function by typing in *Topics in Current Chemistry*.

Color figures are published in full color in the electronic version on SpringerLink.

Aims and Scope

The series *Topics in Current Chemistry* presents critical reviews of the present and future trends in modern chemical research. The scope includes all areas of chemical science, including the interfaces with related disciplines such as biology, medicine, and materials science.

The objective of each thematic volume is to give the non-specialist reader, whether at the university or in industry, a comprehensive overview of an area where new insights of interest to a larger scientific audience are emerging.

Thus each review within the volume critically surveys one aspect of that topic and places it within the context of the volume as a whole. The most significant developments of the last 5–10 years are presented, using selected examples to illustrate the principles discussed. A description of the laboratory procedures involved is often useful to the reader. The coverage is not exhaustive in data, but rather conceptual, concentrating on the methodological thinking that will allow the non-specialist reader to understand the information presented.

Discussion of possible future research directions in the area is welcome.

Review articles for the individual volumes are invited by the volume editors.

In references *Topics in Current Chemistry* is abbreviated *Top Curr Chem* and is cited as a journal.

Impact Factor 2010: 2.067; Section "Chemistry, Multidisciplinary": Rank 44 of 144

Preface: Fast NMR Methods Are Here to Stay

A distinct feature of NMR spectroscopy, the possibility to simultaneously observe hundreds of atoms in complex macromolecules, finds its foundation in the invention of multidimensional experiments almost 40 years ago [1, 2]. The approach, however, has an important caveat: the ultimate resolution obtained in multidimensional experiments comes at a very high price, the long data collection times needed to systematically sample the large multidimensional spectral space. The number of measured data points increases polynomially with the spectrometer field and the desired spectral resolution, and exponentially with the number of dimensions. The problem of lengthy sampling compromises or even prohibits many applications of multidimensional spectroscopy in chemistry and molecular biology. Fortunately, the advent of "fast" NMR spectroscopy offers a number of solutions.

NMR experiments can be dramatically accelerated by reducing the time needed for individual measurements and/or the number of collected data points. Examples of the former include reducing the magnetization recovery time after each scan [3, 4], or spatial encoding of spectral dimensions in single-scan spectroscopy [5, 6]. The contributions to this volume are focused on the latter approach, namely the retrieving of spectral information from a limited number of data points.

The time-consuming systematic sampling of the signal on the entire multidimensional Nyquist grid describing the indirect dimensions is replaced by acquiring FIDs for only a relatively small number of grid points, while preserving all essential information that would be present in the full data set. Two distinct approaches can be traced back to the early years of multidimensional NMR spectroscopy. The former is based on the spectral projection theorem and Fourier Transform [7], and applied for example in the ACCORDION experiment 30 years ago [8]. In the second approach, the positions of the measured points are not constrained and often selected randomly [9]. Both approaches require novel analysis tools and non-standard processing methods, often resulting in significantly increased calculations times, and making them only recently a practical approach.

Deducing three-dimensional information from two-dimensional projections is not a new idea, as pointed out in the first chapter of this volume: the most obvious examples are the three-dimensional descriptions that our brain forms from the two-dimensional images collected by our eyes. The history of projections in high-resolution NMR, from the ACCORDION experiment presented in the early years of multidimensional NMR [8] via reduced dimensionality [10] to GFT [11] and

Projection-Reconstruction [12], has been presented many times, for example [13]. The projection concept for NMR spectroscopy has been implemented in various flavours. In GFT, the following chemical shift combinations for a N-dimensional signal $(\Omega_1, \Omega_2, \ldots \Omega_N)$ spectra are recorded: (Ω_1), $(\Omega_1 \pm \Omega_2)$, $(\Omega_1 \pm \Omega_2 \pm \Omega_3)$, etc. Successive inspection of these spectra yields the chemical shifts in their numbered order. Generalizations of this scheme include the recording of any combination of shifts, e.g. also $(\Omega_1 \pm \Omega_3)$, $(\Omega_2 \pm \Omega_3)$, or the variation of the proportionality factor between the different shifts; the latter is usually referred to as allowing any projection angle ("45°" would correspond to the same number of time increments in all projected dimensions). The restriction to projection angles of "45°" often simplifies the direct interpretation of the projections, where peak picking or reconstructions are deferred to a later stage; an example is multi-way decomposition with PRODECOMP [14]. Reconstruction of the full-dimensional spectrum, for example with the various back-projection schemes implemented in Projection-Reconstruction (Chap. 1), accepts the most general types of projections. The same holds also when each projection is immediately subjected to peak-picking as in the APSY approach [15] (Chap. 2).

Acquiring spectral projections pertains to measuring linear cross sections in the time domain. This can be considered as a special case of a more general sampling scheme, where data points are sampled at any position of the time domain. The method is known as non-uniform or non-linear sampling (NUS or NLS). A historical perspective of this approach is well presented in Chap. 3. It was introduced almost a quarter of a century ago in a seminal publication by Laue and co-workers [9]. In a typical NUS implementation, a small fraction of the data points that would be collected in the conventional uniform sampling is randomly selected and measured. This provides dramatic savings of measurement time. The spectrum is reconstructed using specialized signal processing algorithms such as Maximum Entropy (ME) [9] (see Chaps. 3 and 5), Multi-Dimensional Decomposition (MDD) [16, 17], Fourier transform (FT) [18, 19] (see Chap. 4), Compressed Sensing (CS) [20, 21], etc. The approach provides maximum flexibility in designing the sampling schedule; thus significant efforts in the field are devoted to sampling optimization, which is based on ideas of matched acquisition [9] or improving the random distribution that are used for selecting points for measuring [22].

Despite the fact that fast sampling techniques were known over a long period of time, their broad use by the NMR community started only recently. The turning point was defined by several factors: (1) As a consequence of higher sensitivity provided by a new generation of high-field spectrometers equipped with cryoprobes, the ever-increasing signal frequency range and spectral dimensionality made sampling the limiting condition for more and more practical applications when traditional uniform sampling is used. (2) The increasing demand for high-throughput and automated analysis of an ever-increasing volume of spectral data can only be met by increased resolution and spectra dimensionality. (3) The dramatically increased performance of modern computers makes even the most computationally demanding signal processing algorithms practical. (4) This resulted in the development of novel, powerful algorithms for spectra reconstruction and analysis from sparsely collected measurements.

Within only a few years, fast sampling techniques have been established as an indispensible tool in biomolecular NMR. Sparse sampling is routinely used for resonance assignment and structure determination of globular proteins [23, 24], (Chap. 2), including high-throughput applications by the North-East Structural Genomic Consortium (NESGC) [25, 26] and the Joint Center for Structural Genomics (JCSG) [27].

Spectra of denatured proteins and intrinsically disordered proteins show high peak overlap due to very low dispersions of signal frequencies, making sparse sampling methods a prerequisite for successful analysis. Examples are a 60-residue fragment of nucleoprotein N from the paramyxovirus Sendai [28]; the 148-residue outer membrane protein X (OmpX) from *Escherichia coli* [29]; a 115-residue CD3 Z domain [24]; a 81-residue delta-subunit of RNA polymerase from *Bacillus subtilis* [30]; the 441-residue, intrinsically disordered protein Tau [31]; the 70-residue N-terminal domain of SKIP [32].

The fast sampling has been used in studies of large protein systems: the 86 kDa Maltose-binding protein G [33], a 37 kDa fragment of the *E. coli* enterobactin synthetase module EntF [34], the integral membrane protein Volt-dependent Anion Channel [35] in micelles, the 23 kDa catalytically inactive phosphatase Ssu72 [36], and the 22.4 kD protein kRas [37].

Sparse sampling has been demonstrated also in solid-state NMR [38–41] and metabolomics [42–46].

At this turning moment when novel sampling methods have become routine for resonance assignment and structure determination, we also witness the application of these fast methods to new challenges such as short living molecular systems for example with in-cell NMR [47, 48], unstable proteins, and other cases when measurement time is limited by the sample life time. Integration into automated, comprehensive packages for studies of protein structure and interactions will be one of the next steps of many-fold improving efficiency of biomolecular NMR spectroscopy [49].

This volume presents a discussion of some of the most popular sampling schemes used in "fast" approaches to high-dimensional NMR. Novel ideas, regarding both experimental (sampling) schemes and processing algorithms, keep coming up. In particular, the novel sampling approaches are being integrating with automated assignment, structure determination, and beyond. As the above and many other applications show, "fast" NMR is here to stay.

Göteborg   Martin Billeter and Vladislav Orekhov

References

1. Aue WP, Bartholdi E, Ernst RR (1976) 2-Dimensional spectroscopy – application to nuclear magnetic-resonance. J Chem Phys 64(5):2229–2246
2. Jeener J (1971) Lecture: Proc. AMPERE International Summer School II. Basko Polje, Yugoslavia
3. Pervushin K, Vogeli B, Eletsky A (2002) Longitudinal H-1 relaxation optimization in TROSY NMR spectroscopy. J Am Chem Soc 124(43):12898–12902
4. Schanda P, Kupce E, Brutscher B (2005) SOFAST-HMQC experiments for recording two-dimensional heteronuclear correlation spectra of proteins within a few seconds. J Biomol NMR 33(4):199–211
5. Frydman L, Peng J (1994) Non-Cartesian sampling schemes and the acquisition of 2D NMR correlation spectra from single-scan experiments. Chem Phys Lett 220(4):371–377
6. Frydman L, Scherf T, Lupulescu A (2002) The acquisition of multidimensional NMR spectra within a single scan. Proc Natl Acad Sci U S A 99(25):15858–15862
7. Bracewell RN (1956) Strip integration in radio astronomy. Aust J Phys 9:198–217
8. Bodenhausen G, Ernst RR (1981) The accordion experiment, a simple approach to 3-dimensional NMR spectroscopy. J Magn Reson 45(2):367–373
9. Barna JCJ, Laue ED, Mayger MR, Skilling J, Worrall SJP (1987) Exponential sampling, an alternative method for sampling in two-dimensional NMR experiments. J Magn Reson 73(1):69–77
10. Szyperski T, Wider G, Bushweller JH, Wüthrich K (1993) Reduced dimensionality in triple-resonance NMR experiments. J Am Chem Soc 115(20):9307–9308
11. Kim S, Szyperski T (2003) GFT NMR, a new approach to rapidly obtain precise high-dimensional NMR spectral information. J Am Chem Soc 125(5):1385–1393
12. Kupce E, Freeman R (2003) Projection-reconstruction of three-dimensional NMR spectra. J Am Chem Soc 125(46):13958–13959
13. Malmodin D, Billeter M (2005) High-throughput analysis of protein NMR spectra. Prog Nucl Magn Reson Spectrosc 46(2–3):109–129
14. Malmodin D, Billeter M (2005) Multiway decomposition of NMR spectra with coupled evolution periods. J Am Chem Soc 127(39):13486–13487
15. Hiller S, Fiorito F, Wüthrich K, Wider G (2005) Automated projection spectroscopy (APSY). Proc Natl Acad Sci U S A 102(31):10876–10881
16. Orekhov VY, Ibraghimov I, Billeter M (2003) Optimizing resolution in multidimensional NMR by three-way decomposition. J Biomol NMR 27(2):165–173
17. Orekhov VY, Jaravine VA (2011) Analysis of non-uniformly sampled spectra with multidimensional decomposition. Prog Nucl Magn Reson Spectrosc 59:271–292
18. Coggins BE, Zhou P (2006) Polar Fourier transforms of radially sampled NMR data. J Magn Reson 182(1):84–95
19. Kazimierczuk K, Kozminski W, Zhukov I (2006) Two-dimensional Fourier transform of arbitrarily sampled NMR data sets. J Magn Reson 179(2):323–328
20. Holland DJ, Bostock MJ, Gladden LF, Nietlispach D (2011) Fast multidimensional NMR spectroscopy using compressed sensing. Angew Chem-Int Edit 50(29):6548–6551
21. Kazimierczuk K, Orekhov VY (2011) Accelerated NMR spectroscopy by using compressed sensing. Angew Chem-Int Edit 50(24):5556–5559
22. Hyberts SG, Takeuchi K, Wagner G (2010) Poisson-gap sampling and forward maximum entropy reconstruction for enhancing the resolution and sensitivity of protein NMR data. J Am Chem Soc 132(7): 2145–+
23. Fiorito F, Hiller S, Wider G, Wüthrich K (2006) Automated resonance assignment of proteins: 6D APSY-NMR. J Biomol NMR 35(1):27–37
24. Jaravine VA, Zhuravleva AV, Permi P, Ibraghimov I, Orekhov VY (2008) Hyperdimensional NMR spectroscopy with nonlinear sampling. J Am Chem Soc 130(12):3927–3936

25. Lemak A, Gutmanas A, Chitayat S, Karra M, Fares C, Sunnerhagen M, Arrowsmith CH (2011) A novel strategy for NMR resonance assignment and protein structure determination. J Biomol NMR 49(1):27–38
26. Liu G, Shen Y, Atreya HS, Parish D, Shao Y, Sukumaran DK, Xiao R, Yee A, Lemak A, Bhattacharya A, Acton TA, Arrowsmith CH, Montelione GT, Szyperski T (2005) NMR data collection and analysis protocol for high-throughput protein structure determination. Proc Natl Acad Sci USA 102(30):10487–92
27. Wüthrich K (2010) NMR in a crystallography-based high-throughput protein structure-determination environment. Acta Crystallogr F-Struct Biol Cryst Commun 66:1365–1366
28. Pannetier N, Houben K, Blanchard L, Marion D (2007) Optimized 3D-NMR sampling for resonance assignment of partially unfolded proteins. J Magn Reson 186(1):142–149
29. Hiller S, Wasmer C, Wider G, Wüthrich K (2007) Sequence-specific resonance assignment of soluble nonglobular proteins by 7D APSY-NMR spectroscopy. J Am Chem Soc 129(35):10823–10828
30. Motackova V, Novacek J, Zawadzka-Kazimierczuk A, Kazimierczuk K, Zidek L, Anderova HS, Krasny L, Kozminski W, Sklenar V (2010) Strategy for complete NMR assignment of disordered proteins with highly repetitive sequences based on resolution-enhanced 5D experiments. J Biomol NMR 48(3):169–177
31. Narayanan RL, Durr UHN, Bibow S, Biernat J, Mandelkow E, Zweckstetter M (2010) Automatic assignment of the intrinsically disordered protein Tau with 441-residues. J Am Chem Soc 132(34):11906–11907
32. Wen J, Wu JH, Zhou P (2011) Sparsely sampled high-resolution 4-D experiments for efficient backbone resonance assignment of disordered proteins. J Magn Reson 209(1):94–100
33. Tugarinov V, Choy WY, Orekhov VY, Kay LE (2005) Solution NMR-derived global fold of a monomeric 82-kDa enzyme. Proc Natl Acad Sci USA 102(3):622–627
34. Frueh DP, Sun ZYJ, Vosburg DA, Walsh CT, Hoch JC, Wagner G (2006) Non-uniformly sampled double-TROSY hNcaNH experiments for NMR sequential assignments of large proteins. J Am Chem Soc 128(17):5757–5763
35. Hiller S, Garces RG, Malia TJ, Orekhov VY, Colombini M, Wagner G (2008) Solution structure of the integral human membrane protein VDAC-1 in detergent micelles. Science 321(5893):1206–1210
36. Werner-Allen JW, Coggins BE, Zhou P (2010) Fast acquisition of high resolution 4-D amide-amide NOESY with diagonal suppression, sparse sampling and FFT-CLEAN. J Magn Reson 204(1):173–178
37. Gossert AD, Hiller S, Fernandez C (2011) Automated NMR resonance assignment of large proteins for protein-ligand interaction studies. J Am Chem Soc 133(2):210–213
38. Huber M, Hiller S, Schanda P, Ernst M, Bockmann A, Verel R, Meier BH (2011) A proton-detected 4D solid-state NMR experiment for protein structure determination. ChemPhysChem 12(5):915–918
39. Jones DH, Opella SJ (2006) Application of maximum entropy reconstruction to PISEMA spectra. J Magn Reson 179(1):105–113
40. Matsuki Y, Eddy MT, Griffin RG, Herzfeld J (2010) Rapid three-dimensional MAS NMR spectroscopy at critical sensitivity. Angew Chem-Int Edit 49(48):9215–9218
41. Rovnyak D, Filip C, Itin B, Stern AS, Wagner G, Griffin RG, Hoch JC (2003) Multiple-quantum magic-angle spinning spectroscopy using nonlinear sampling. J Magn Reson 161(1):43–55
42. Bingol K, Bruschweiler R (2011) Deconvolution of chemical mixtures with high complexity by NMR consensus trace clustering. Anal Chem 83(19):7412–7417
43. Hyberts SG, Heffron GJ, Tarragona NG, Solanky K, Edmonds KA, Luithardt H, Fejzo J, Chorev M, Aktas H, Colson K, Falchuk KH, Halperin JA, Wagner G (2007) Ultrahigh-resolution H-1-C-13 HSQC spectra of metabolite mixtures using nonlinear sampling and forward maximum entropy reconstruction. J Am Chem Soc 129(16):5108–5116

44. Pontoizeau C, Herrmann T, Toulhoat P, Elena-Herrmann B, Emsley L (2010) Targeted projection NMR spectroscopy for unambiguous metabolic profiling of complex mixtures. Magn Reson Chem 48(9):727–733
45. Zhang FL, Bruschweiler-Li L, Bruschweiler R (2010) Simultaneous de novo identification of molecules in chemical mixtures by doubly indirect covariance NMR spectroscopy. J Am Chem Soc 132(47):16922–16927
46. Zhang FL, Robinette SL, Bruschweiler-Li L, Bruschweiler R (2009) Web server suite for complex mixture analysis by covariance NMR. Magn Reson Chem 47:S118–S122
47. Sakakibara D, Sasaki A, Ikeya T, Hamatsu J, Hanashima T, Mishima M, Yoshimasu M, Hayashi N, Mikawa T, Walchli M, Smith BO, Shirakawa M, Guntert P, Ito Y (2009) Protein structure determination in living cells by in-cell NMR spectroscopy. Nature 458(7234):102–U10
48. Selenko P, Frueh DP, Elsaesser SJ, Haas W, Gygi SP, Wagner G (2008) In situ observation of protein phosphorylation by high-resolution NMR spectroscopy. Nat Struct Mol Biol 15(3):321–329
49. Billeter M, Wagner G, Wüthrich K (2008) Solution NMR structure determination of proteins revisited. J Biomol NMR 42(3):155–158

Contents

Concepts in Projection-Reconstruction 1
Ray Freeman and Ēriks Kupče

Automated Projection Spectroscopy and Its Applications 21
Sebastian Hiller and Gerhard Wider

**Data Sampling in Multidimensional NMR: Fundamentals
and Strategies** 49
Mark W. Maciejewski, Mehdi Mobli, Adam D. Schuyler, Alan S. Stern,
and Jeffrey C. Hoch

Generalized Fourier Transform for Non-Uniform Sampled Data 79
Krzysztof Kazimierczuk, Maria Misiak, Jan Stanek,
Anna Zawadzka-Kazimierczuk, and Wiktor Koźmiński

Applications of Non-Uniform Sampling and Processing 125
Sven G. Hyberts, Haribabu Arthanari, and Gerhard Wagner

**Erratum to: Data Sampling in Multidimensional NMR: Fundamentals
and Strategies** 149
Mark W. Maciejewski, Mehdi Mobli, Adam D. Schuyler,
Alan S. Stern, and Jeffrey C. Hoch

Index 151

Concepts in Projection-Reconstruction

Ray Freeman and Ēriks Kupče

Abstract The Achilles heel of conventional multidimensional NMR spectroscopy is the long duration of the measurements, set by the Nyquist sampling condition and the resolution requirements in the evolution dimensions. Projection-reconstruction solves this problem by radial sampling of the evolution-domain signals, relying on Bracewell's Fourier transform slice/projection theorem to generate a set of projections at different inclinations. Reconstruction is implemented by one of three possible deterministic back-projection schemes (additive, lowest-value, or algebraic), or by a statistical model-fitting program. For simplicity the treatment focuses principally on the three-dimensional case, and then extends the analysis to four dimensions. The concept of hyperdimensional spectroscopy is described for dealing with even higher dimensions.

Keywords Bayesian · Correlation · Deterministic · Hyperdimensional · Multidimensional · Nuclear magnetic resonance · Projection-reconstruction · Sparse sampling

Contents

1 Introduction 2
2 Three-Dimensional NMR 3
3 Reconstruction 6
  3.1 Deterministic Reconstruction 6
  3.2 Statistical Methods 13
4 Four-Dimensional Spectroscopy 15
5 Hyperdimensional Spectroscopy 17
6 Conclusions 18
References 19

R. Freeman (✉)
Jesus College, Cambridge University, Cambridge, UK
e-mail: rf110@hermes.cam.ac.uk

Ē. Kupče
Agilent Technologies, 6 Mead Road, Yarnton, Oxford, UK

1 Introduction

We live in a three-dimensional world. Survival has ensured that our brains have evolved a remarkable capacity to reconstruct a three-dimensional image based on a pair of slightly different two-dimensional views of our environment. While we take this apparent instance of projection-reconstruction entirely for granted, the complexity of the general problem soon becomes apparent in the science of robotics, when we attempt to teach a machine to construct a reliable visual model of its surroundings. What for humans is an entirely automatic process needs to be derived again from first principles for a perambulating robot. Concepts like *parallax* and *occultation* have to be re-examined.

Art presents similar challenges. It was quite some time before artists discovered *perspective* – the key to depicting a plausible representation of three dimensions on a plane canvas. Nowhere is this challenge more critical than in sculpture, the creation of three-dimensional artefacts that reconcile the visual and tactile senses. It is reported that the famous French sculptor Auguste Rodin employed an unusual stratagem – he placed his model on a turntable with strong back-lighting, concentrating his attention on the changing silhouettes as he rotated the table in small steps. This scheme was certainly effective; his sculptures of the human form are so life-like that his critics accused him of cheating by taking plaster casts of his subjects. A good case can be made that Rodin was the true father of projection-reconstruction.

Present-day computers greatly facilitate this transformation of two-dimensional raw data into a three-dimensional image. Artefacts in a museum collection are often irreplaceable, so that worldwide dissemination is quite impractical. However, if a sequence of digital photographs is taken from several different points of view, a program can be written to reconstruct an image that can be rotated about an axis to give a lifelike representation of three spatial dimensions [1]. The resulting digital archive is readily transferable to any desired location and can be scaled if necessary. A more mundane application is to offer some article for sale on a popular website in a form that conveys a three-dimensional impression. Google Earth offers street views of many cities that give the perception of three-dimensional reality, while more recent research [1] creates a true three-dimensional reconstruction of the local street environment. Lost in a strange city, a person could in principle use a mobile phone to photograph an adjacent building, pass on the information to a distant computer, and receive confirmation of his present location (within a metre) and also his orientation, followed by detailed directions for proceeding to his intended destination. Face-recognition software employs related procedures to track a person in a crowd.

X-Ray tomography [2] makes use of the same principles. A set of pictures of X-ray absorption is taken at different angles of incidence around a circle. The software uses this information to reconstruct an image of the internal organs of the patient. Whereas traditional X-ray studies gave only a two-dimensional image on a sheet of photographic film, tomography allows the surgeon to examine the internal structure as if in three spatial dimensions. At about the same time, and completely independently, Paul Lauterbur [3] hit on the idea of medical imaging by recording nuclear magnetic resonance absorption in an applied magnetic field gradient. By combining the results of measurements at different inclinations of the magnetic field gradient, he was able to reconstruct a map of the distribution of protons within the sample. Soon the 'sample' became a human patient and the exciting science of MRI was born.

Projection-reconstruction is not therefore a new phenomenon. Recently it has become of particular interest to high-resolution NMR spectroscopists with the realization [4] that a three-dimensional spectrum can be treated as a candidate for reconstruction in just the same manner as a physiological sample, but with the advantage that the 'object' is now a sparse distribution of discrete resonances, like the stars in the night sky, not a continuous absorption medium. (Interestingly, projection-reconstruction borrows at least two data processing schemes from earlier work in radio astronomy.) Sparse sampling assumes particular importance as spectra are recorded in higher and higher dimensions in order to study larger and larger biomolecules, often with isotopic enrichment in both carbon-13 and nitrogen-15. The prime concern is speed. This review focuses on data-sampling methodology rather than the actual spectroscopic applications.

2 Three-Dimensional NMR

The basic principles of projection-reconstruction are most easily understood by reference to the simplest case – three dimensional spectroscopy. Early experiments in NMR were preoccupied with the inherently poor sensitivity. The duration of a measurement was often dictated by the need for appreciable multiscan averaging. On the other hand, multidimensional spectra must normally satisfy the Nyquist sampling condition and the resolution requirements in each and every evolution dimension, so the number of scans is inevitably large. In modern spectrometers, particularly those equipped with a cryogenically cooled probe (receiver coil and preamplifier), this usually ensures a satisfactory signal-to-noise ratio long before all the evolution dimensions have been explored on a full Cartesian matrix. The measurement duration is said to be 'sampling-limited' rather than 'sensitivity-limited'. The obvious remedy is to resort to some form of sparse sampling of evolution space.

Sacrifices must therefore be made. All sparse sampling regimes come at the expense of spectral artefacts. Some introduce an element of randomness in the selection of sampling co-ordinates, but although this can reduce the mean intensity of artefacts, it does so at the expense of widespread proliferation. Better the devil you know. Of the many possible schemes, radial sampling appears to offer the most acceptable solution. Because the resulting artefacts are well defined, effective suppression schemes can be devised. More important is the fact that radial sampling in the time domain gives rise to a particularly simple observable result – projections of the target spectrum in the frequency domain.

The key 'slice/projection' theorem was first formulated in a radio astronomy context by Bracewell [5] and later exploited in NMR by Nagayama et al. [6] and Bodenhausen and Ernst [7, 8]. Consider the case of a typical plane $S(F_1,F_2)$ from a three-dimensional NMR spectrum $S(F_1,F_2,F_3)$. In order to obtain a projection at some angle α, the theorem postulates that the time domain response should be sampled along a slice through the origin at this same angle α. This requires that the evolution parameters t_1 and t_2 be varied jointly [7–13]:

$$t_1 = t \cos\alpha, \tag{1}$$

$$t_2 = t \sin\alpha. \tag{2}$$

Fourier transformation of this skew slice through two-dimensional evolution space provides the required projection (Fig. 1).

Suppose that the NMR signal from a typical chemical site evolves at a frequency Ω_A in t_1, and one from a second correlated site evolves at Ω_B during t_2. The signal component that is observed after the second evolution stage is modulated as $\cos(\Omega_A\, t_1)\cos(\Omega_B\, t_2)$. As in the standard practice, quadrature detection is employed in both evolution intervals, generating four signal components:

$$S_1 = \cos(\Omega_A\, t_1)\cos(\Omega_B\, t_2), \tag{3}$$

$$S_2 = \sin(\Omega_A\, t_1)\cos(\Omega_B\, t_2), \tag{4}$$

$$S_3 = \cos(\Omega_A\, t_1)\sin(\Omega_B\, t_2), \tag{5}$$

$$S_4 = \sin(\Omega_A\, t_1)\sin(\Omega_B\, t_2). \tag{6}$$

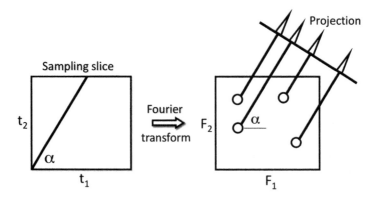

Fig. 1 The Bracewell slice/projection theorem. The Fourier transform of a slice through the evolution dimension at an inclination α (*left*) is the projection of the corresponding frequency-domain spectrum at the same angle α (*right*)

After substitution of (1) and (2) the appropriate combinations of these terms creates the four signals:

$$S_1 - S_4 = \cos(\Omega_A t \cos\alpha + \Omega_B t \sin\alpha), \tag{7}$$

$$S_2 + S_3 = \sin(\Omega_A t \cos\alpha + \Omega_B t \sin\alpha), \tag{8}$$

$$S_1 + S_4 = \cos(\Omega_A t \cos\alpha - \Omega_B t \sin\alpha), \tag{9}$$

$$S_2 - S_3 = \sin(\Omega_A t \cos\alpha - \Omega_B t \sin\alpha). \tag{10}$$

Hypercomplex Fourier transformation gives the sum and difference frequencies (scaled accordingly) given by ($\Omega_A\cos\alpha + \Omega_B\sin\alpha$) and ($\Omega_A\cos\alpha - \Omega_B\sin\alpha$). Consequently each measurement produces a pair of projections inclined at $\pm\alpha$. Figure 2 shows five projections of a simulated two-dimensional spectrum containing seven peaks. They were calculated as integrals of the intensities along rays perpendicular to the projected trace. In the time domain the angle α must be positive whereas in the frequency domain α can take on all angles 0° to 360°, although the projections at α and $\alpha \pm 180°$ are of course identical.

Because the number of time-domain slices (and hence the number of recorded projections) is relatively small, the density of sampling points is far lower than the density used in the conventional experiment, which must examine every point on the complete Cartesian grid while satisfying the Nyquist condition and the requirement for adequate resolution. This is where the critical time saving occurs. With this limited radial sampling [13], the speed advantage increases by an order of magnitude for each new evolution dimension beyond the first. This opens up the

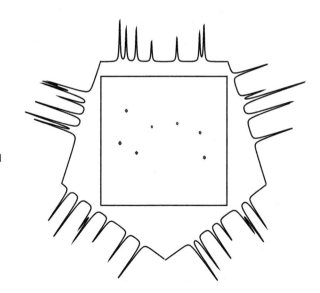

Fig. 2 A set of five integral projections of a simulated two-dimensional spectrum containing seven responses. Note the situation (*upper right*) where two responses are eclipsed, giving a projected response with an increased integral in that particular direction

possibility of studying unstable molecules, or chemically exchanging systems, or even some protein folding applications. Naturally the sensitivity falls off as the measurement duration is reduced, but it is assumed here that sensitivity is not a limiting factor.

3 Reconstruction

Projections are therefore relatively easily obtained, but the following reconstruction stage is more challenging. Formally this involves the inverse Radon transform [14, 15] – computing the three-dimensional spectrum $S(F_1,F_2,F_3)$ starting from all the recorded projections. Inverse problems of this kind are notoriously tricky to solve but an NMR spectrum is a favourable case because the target spectrum comprises discrete resonances sparsely distributed in three dimensions rather than a continuum of absorption. There are two general approaches to this problem – deterministic and statistical [16].

3.1 Deterministic Reconstruction

The full three-dimensional spectrum $S(F_1,F_2,F_3)$ is built up by assembling individual reconstructed planes $S(F_1,F_2)$ as a function of the directly detected dimension F_3. The basic procedure for reconstructing $S(F_1,F_2)$ is best described as 'back-projection'. Suppose that there are n one-dimensional projection traces available for the reconstruction. Consider a typical trace P_1, recorded at some arbitrary angle α. Every peak in P_1 is extended at right angles to the trace to form a set of parallel ridges running across the plane $S(F_1,F_2)$. These ridges have cross-sections defined by the resonance lineshapes in P_1. Another set of ridges from a differently oriented projection trace P_2 intersect with those from P_1, and the point of intersection defines the location of a potential correlation peak of the target spectrum. If the signals are added there is a peak at the point of intersection (Fig. 3). A set of P_n back-projections is measured. Usually these include $\alpha = 0°$ and $\alpha = 90°$ projections, obtained by Fourier transformation of time-domain signals recorded with $t_2 = 0$ or $t_1 = 0$, because they have a relatively high sensitivity [17]. When all the projections are combined, the genuine correlation peaks become better defined in comparison with the artefacts.

The ambiguity between genuine and false correlations can normally be resolved in terms of the number of back-projected ridges that intersect at the same location. A genuine correlation peak involves the intersection of n ridges – *one from each and every trace P_n*. Intersections involving less than n ridges can normally be taken to indicate false correlations, although in practice this criterion may not be entirely clear cut, notably in situations where some projections contain very

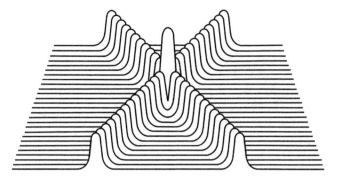

Fig. 3 The intersection of two back-projected ridges in the additive mode creates a correlation peak but leaves undesirable ridges

weak or missing resonances. Exactly how the intersecting ridges are combined is determined by the back-projection algorithm used [18]. There are three principal methods for reconstructing the two-dimensional spectrum $S(F_1,F_2)$ by combining back-projections. Each approach has strengths and disadvantages, and the choice is mainly determined by the nature of the available experimental projection data.

3.1.1 The Additive Algorithm

Consider a typical pixel in the $S(F_1,F_2)$ plane. If it corresponds to a genuine correlation response there are n signal-bearing rays intersecting at that point, one from each of the n projections. The simplest procedure is to add these n contributions to signal intensity of this pixel (or alternatively, calculate the arithmetic mean). This has the advantage that all n traces contribute to the final signal-to-noise ratio, just as in multiscan averaging. Figure 4 illustrates the improvement in spectral quality as the number of measured projections n is increased from 6 through 18. Not only does the signal-to-noise ratio increase, but artefacts also become less apparent, indicating that increasing n is to be preferred over time-averaging identical traces. One advantage is that the additive algorithm allows for the possibility that some projection traces may be missing a particular resonance through poor sensitivity; genuine correlation peaks then occur at lower-order ($<n$) intersections. Even in the case where there is a very noisy projection trace with no detectable signals, the additive reconstruction remains valid.

The additive algorithm has the advantage of being linear. The correct relative intensities are normally preserved, and there is no spurious 'improvement' in the signal-to-noise ratio in the reconstructed spectrum; the noise floor increases as the square root of n. This algorithm proves to be most useful when n is relatively large, for then the intensities of residual ridges and false cross-peaks are weak in comparison with the genuine correlation peaks, and may fall below the general level of the noise. However, the presence of many vestigial ridges in the skirts of a reconstructed resonance distorts and broadens the derived line-shape. This can be

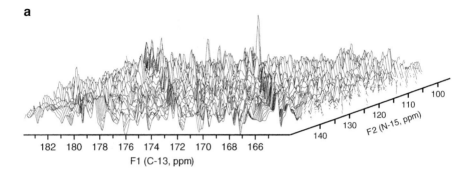

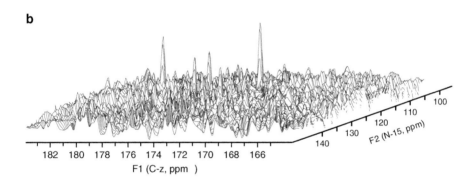

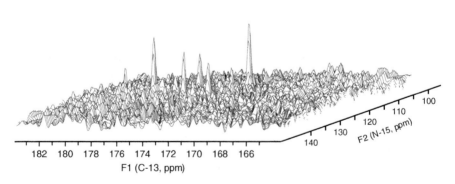

Fig. 4 Reconstructed spectra using the additive back-projection algorithm, showing the effect of increasing the number of projections, (**a**) $n = 6$, (**b**) $n = 12$, (**c**) $n = 18$. Six responses are detected as the signal-to-noise ratio increases and the artefacts become less obtrusive. Residual ridges are apparent in (**a**) but not in (**c**)

corrected by the application of a resolution enhancement function to the projection traces [19] – a procedure known as filtered back-projection.

If necessary, a reprocessing program can be written to filter out artefacts on the grounds that each true correlation peak carries with it a known, well-defined pattern of back-projection ridges. One such program, 'CLEAN', has been adapted from a procedure first introduced in radio astronomy [20, 21] and later applied in NMR spectroscopy [22, 23]. It presupposes that the line shapes in the reconstructed spectrum are known, or can be measured. Then an iterative search selects the tallest response in the reconstructed $S(F_1,F_2)$ plane and subtracts it, along with its associated back-projection ridges, storing the appropriate intensity and frequency co-ordinates in a table. The next iteration stage is slightly less burdened with artefactual ridges, and the next-tallest response is located and removed, along with its associated ridges. The procedure continues until the detection threshold is just above the base-plane noise, where further iteration becomes unproductive. The only remaining danger is that extremely weak NMR responses, comparable with the baseline noise, could be overlooked. The spectroscopist may then make direct use of the correlation information stored in the table, or alternatively, reconstruct a processed version of the spectrum.

3.1.2 The Lowest-Value Algorithm

In situations where artefacts are of more serious concern than any sensitivity considerations, a more appropriate approach is to superimpose all n back-projection rays, but retain only the *lowest amplitude* at each pixel [24, 25]. (More specifically, the program selects the lowest absolute magnitude response and then reinstates the original sign.) Then the only intersections that give rise to correlation peaks are those involving one ridge from each trace. Intersections of less than n ridges necessarily overlap with noise from a back-projection that carries no NMR signal, causing any *potential* false correlation peak to be replaced by the base-plane noise. A similar suppression occurs for all the extraneous ridges. For this reason the lowest-value algorithm generates a very clean reconstruction, because each additional back-projection operation constrains the artefacts more effectively. Sensitivity does not improve with n; indeed it is determined by the signal-to-noise ratio of the weakest resonance in one of the projections. Consequently the reconstruction process breaks down if one of the n traces has a missing resonance, unless this eventuality is recognized and that particular projection is deliberately eliminated from the reconstruction. The lowest-value algorithm is most useful when n is small. Its inherent non-linearity has two consequences. First, the skirts of the reconstructed peaks are clipped; instead of circular (or elliptical) intensity contours, some polyhedral character is imposed, with $2n$ edges. Second, the *character* of the base-plane noise is changed, because the algorithm selects the lowest noise amplitude at each pixel.

However, the principal restriction of the lowest-value algorithm is to experiments where *all* projections have an acceptable signal-to-noise ratio. Otherwise problems arise because the algorithm discriminates against very weak signals comparable with the noise. One bad apple spoils the whole barrel. At any given pixel, the intensity is set by the one particular back-projected ray that just happens to contribute a near-zero noise fluctuation. In this situation the degree of signal suppression may vary from pixel to pixel across the region where a weak correlation peak is expected, leading to break up of the peak profile.

3.1.3 Hybrid Schemes

One solution is to devise a methodology that combines the advantages of the additive and lowest value procedures while avoiding the pitfalls of each. These hybrid schemes seek to balance the advantages of *accumulation* and *purging*. An initial accumulation stage combines the reconstructed spectra in the additive mode to improve sensitivity, and then any artefacts are purged by the lowest-value algorithm. The n experimental projections are divided into p independent batches, usually of equal size k. These subsets are used to reconstruct p different versions of the desired $S(F_1,F_2)$ spectrum, enhanced in signal-to-noise ratio by applying the additive algorithm to each batch in turn. Then the resulting p reconstructions are combined pixel-by-pixel according to the lowest-value algorithm in order to minimize the artefacts. The ratio k:p determines the balance between the conflicting demands of sensitivity and artefact suppression. In the limit that $k = n$ this regime reduces to pure accumulation; at the opposite extreme where $p = n$ this scheme reduces to pure purging. This simple hybrid algorithm is effective, but is not necessarily the optimum scheme.

In an attempt to improve on this hybrid scheme, a combinatorial approach has been suggested [26]. Instead of accumulating p *independent* batches, this procedure examines a much larger number of batches $^nC_k = n!/(n-k)!k!$, representing the sums of all possible combinations of k amplitudes chosen from the available total n. The perceived rationale for this combinatorial method is that processing an *extremely* high number of batches should deliver a substantial sensitivity advantage as the lowest-value operation is applied to all ^nC_k sums. Since these calculations must be repeated for every pixel in the $S(F_1,F_2)$ plane, and for every plane as a function of F_3, the method is very computationally intensive.

Mandelshtam [27] has suggested a fast and very effective simplification. When *all possible* combinations of n signals in batches of k have been examined to search for the batch with the lowest sum, all these low-intensity items *must necessarily* be found in one particular batch, so the result is simply the sum of the k lowest-amplitude signals. The slow combinatorial calculation can therefore be replaced by a single, fast summation. All n back-projected rays that intersect at a given pixel are examined, and the subset with the k lowest amplitudes is retained. As before, the adjustable parameter k serves to define the desired balance between sensitivity and artefact suppression. Clearly this is the most effective hybrid scheme discovered so far.

3.1.4 The Algebraic Algorithm

When the aim is to study the very crowded spectra characteristic of large biomolecules such as proteins, the overriding concern is to reduce the amount of data to be processed. Then it makes sense to simplify the information in the raw projection traces by eliminating all except the frequency information, ignoring intensities and line shapes. Each projection trace is processed with a peak-picking routine and all further processing is based *solely* on frequency information. The n projection traces are replaced by n lists of frequencies, each list from a different projection at a specific projection angle α. Here lies the real danger – peak-picking can miss a resonance because it is lying on the shoulder of a stronger response, or simply because a weak resonance lies below the arbitrary intensity threshold assumed by the peak-picking program. Then the criterion for recognizing a genuine correlation peak (n intersections) is compromised.

Apart from these caveats, the algorithm is an exercise in simple algebra [15, 28]. It selects one frequency from each of the n lists, thus defining n intersecting straight lines running across the reconstruction plane $S(F_1,F_2)$ at various inclinations α. By solving the resulting n simultaneous equations, the algorithm determines whether or not *all* these straight lines intersect at a point. In practice a certain degree of leeway is allowed, based on the expected accuracy of the frequency measurements. The process is continued until a positive outcome is detected – all n straight lines meet within a small, predefined 'area of uncertainty'. The 'centre of gravity' is taken as the location of a genuine correlation peak. The frequency co-ordinates used in these n-fold solutions are then saved, and the corresponding frequencies removed from the frequency lists.

Problems arise because there may still be further *genuine* correlations that involve less than n intersecting lines, owing to the shortcomings of the peak-picking routine. An iterative scheme is contrived [29] which relaxes the requirement that there be n intersections. The next stage examines combinations of frequencies from the depleted projection lists, accepting all $(n-1)$-fold intersections as genuine correlations, and transferring the corresponding frequency co-ordinates to the 'accepted' list. There is no *absolute* guarantee that these new solutions do not contain an occasional false correlation, but the probability is minimized because a large number of resonance frequencies associated with n-fold solutions has already been removed from the list. A third level of iteration may then be initiated, searching for possible genuine solutions of the order $(n-2)$ and so on, until the operator terminates the search.

The power of this algebraic algorithm stems from the very high degree of data reduction achieved in the peak-picking stage, something that is indispensable when dealing with spectra of very high complexity. A possible disadvantage is the lack of a cast-iron criterion for identifying false correlations, but the saving grace is that for large biomolecules an absolutely complete solution may not be necessary. Note that although this mode of operation is correctly categorized as back-projection, it does not involve 'reconstruction' in the spectroscopic sense. Correlations appear as frequency co-ordinates in the 'accepted' list, with no opportunity for viewing a reconstructed spectrum to make a judgement about reliability. This apparently clean end-result is illusory because information about signal intensities compared with levels of artefacts and noise has been intentionally disregarded. The method has been applied successfully to multidimensional spectra of proteins [29].

3.1.5 Eclipsed Resonances

Complications arise whenever two responses are eclipsed – where the projection has been recorded at an inclination α that happens to catch two peaks in the $S(F_1,F_2)$ plane in exact alignment. Consider first of all the common case where all responses are positive. One example is illustrated in Fig. 2 (upper right). In the additive algorithm, back-projection then makes a twofold contribution to the intensities at *both* locations, distorting the relative intensities in the final reconstruction. Fortunately the severity of this intensity error decreases with n. One remedy is to discard the highest and lowest back-projected contributions to the intensity of a given pixel on the grounds that they could be unreliable, then sum the rest. By its very nature the lowest-value algorithm is more forgiving when there are eclipsed peaks; an abnormally intense response in one projection is unlikely to affect the corresponding pixel. Because the algebraic algorithm retains no intensity information but relies solely on frequencies, it is essentially unaffected by eclipsed back-projection.

There is a far more serious problem when the $S(F_1,F_2)$ spectrum is composed of both positive and negative resonances, since the eclipsed condition can lead to cancellation (or severe attenuation) of the corresponding projected signal. This interference between signals of opposite phase affects the three back-projection algorithms in quite different ways. The additive scheme (with n large) should tolerate occasional cancellation effects reasonably well, since if one direction of back-projection proves ineffective this does little to falsify the overall reconstruction. The lowest-value algorithm is far more sensitive to accidental cancellation because destructive interference can seriously degrade the reconstruction.

A comprehensive solution to interference between eclipsed resonances is provided by a subroutine that sets up the radial sampling in such a way as to avoid all those projection directions α that would lead to eclipsed peaks [18]. It is based on initial sampling with $t_2 = 0$ or $t_1 = 0$, generating the 0° and 90° projections after Fourier transformation. Two-dimensional convolution of the responses along these axes produces a *preliminary* test map containing both genuine and false correlation peaks. Projections that record not the integral, but the tallest signal on the projection ray are known as 'skyline projections'. These are computed at all possible angles α, and the integral over each projection trace is plotted as a function of α. This graph displays a constant integral except at inclinations where two or more responses are eclipsed, when there is a sudden dip. The graph overestimates the danger that genuine peaks are eclipsed because false correlation peaks make contributions to the projections, but it can nevertheless be used to predict those projection angles that avoid all possible cases of overlap.

There are alternative strategies for treating spectra with positive and negative responses. They work best if the projection angles are chosen to avoid eclipsed peaks. One method divides the projection information into two independent sets – one with positive signals and the other with negative signals. These *plus* and *minus* sets are used separately to reconstruct *plus* and *minus* $S(F_1,F_2)$ planes, which are then recombined. Another scheme converts all the resonances in the projections into positive peaks, thereby limiting the reconstruction to the absolute magnitude mode.

Mandelshtam [27] has proposed a histogram-based algorithm for reconstructing spectra with both positive and negative peaks, and it does not rely on avoiding the eclipsed case. It retains only the most-likely contribution to the intensity at a given pixel. An artificially broadened amplitude distribution function is derived from the histogram representing all the intensity contributions. Although loosely related to the sum of the individual amplitudes, the maximum of this function is quite insensitive to cancellation effects. This scheme works better at higher values of n. It has been successfully tested on simulated two-dimensional spectra.

3.1.6 Projected Linewidths

In a three-dimensional experiment it is quite likely that the nuclei evolving in t_1 and t_2 have different spin–spin relaxation times T_2^A and T_2^B. This means that a response in the $S(F_1,F_2)$ plane may have very different natural linewidths in the two frequency dimensions. With a skew slice through evolution space at an angle α, Fourier transformation generates a projected response with a Lorentzian width given by

$$\Delta v = \cos\alpha/(\pi T_2^A) + \sin\alpha/(\pi T_2^B). \tag{11}$$

This response is broader than at least one of the parent lines in F_1 or F_2. This may suggest a choice of projection angle α that favours a narrower projected line if good resolution is an important consideration.

3.2 *Statistical Methods*

An entirely different approach to reconstruction [16] is to find a model of the two-dimensional spectrum $S(F_1,F_2)$ that is compatible with all the measured projection traces. In principle the iteration could start with an arbitrary or completely featureless model (zero intensity at every pixel), but usually it is better to employ some 'prior knowledge'. In the vicinity of a correlation peak it is clear that there must be some correlation between the intensities of adjacent pixels. Prior knowledge may take the form of *assumptions* about lineshapes or the expected number of resonances in the two-dimensional spectrum, or it might exploit hard evidence from an earlier deterministic scheme. At the most primitive level, where each pixel in the $S(F_1,F_2)$ plane is fitted *independently*, these statistical programs converge very slowly, but there is much to be gained by restricting the variable parameters to a number of discrete resonances with appropriate line-shapes, for example two-dimensional Gaussians. There is an inherent danger in these assumptions because a spurious spike in the noise could be 'promoted' to the status of a genuine correlation peak – this particular wolf has been provided with sheep's clothing. A standard least-squares procedure may be used, but convergence to a global solution is faster if the more sophisticated simulated annealing routine [30] is employed. The 'maximum likelihood' estimate [31], loosely related to least-squares fitting, seeks to maximize the probability of observing the set of experimental projections P_n given the current proposed $S(F_1, F_2)$ map.

Maximum entropy reconstruction [32] is claimed to return a 'maximally non-committal' solution. It calculates a small set of proposed $S(F_1, F_2)$ maps that are compatible with the measured projections within the experimental errors, and selects the one with the least information content. For this reason it suppresses all noise and artefacts in the reconstruction and is therefore prone to be misleading. In another terminology, it rejects 'false positives' but is likely to return 'false negatives'. This particular feature suggests that the maximum entropy solution could prove to be a useful starting point for more sophisticated statistical programs.

3.2.1 Bayesian Inference [33]

This is a learning system that tests the degree to which the suggested model 'M' is consistent with the experimental data 'D', and any prior knowledge about the problem 'C'. It proposes an initial model two-dimensional spectrum $S(F_1, F_2)$ in the light of any prior assumptions, for example the expected lineshapes. This defines a *conditional* prior probability $P(M|C)$ that the model is correct based only on the initial assumptions. The next stage updates $P(M|C)$ in the light of the experimental projection data D to give the *posterior* probability $P(M|DC)$ reflecting how well the proposed model is justified based on both D and C. The next parameter is the likelihood $P(D|MC)$ that the experimental data D is consistent with the model M and the prior assumptions C. Bayes' theorem can be expressed as

$$P(M|DC) \propto P(M|C)P(D|MC). \qquad (12)$$

It is now possible to maximize the posterior probability $P(M|DC)$ to give the most probable model for the two-dimensional spectrum $S(F_1, F_2)$. There are many methods available for such a computation, including the Markov chain Monte-Carlo algorithm.

3.2.2 The Markov Chain Monte-Carlo Method

Monte Carlo methods originated in an ingenious approach to the complex problem of evaluating the probability that a hand of Solitaire would come out successfully.

The solution was to set out several Solitaire hands at random and count the proportion of successful hands. A Markov chain defines a sequence of states where the 'transition probability' from the current state of a system to its next state is dependent only on the value of the current state. The starting point can be arbitrary but the chain must eventually reach a stationary distribution. It should not get trapped in a loop, and must also retain some probability of jumping to the next state. To avoid bias in the choice of starting conditions the initial set of results is usually discarded, a procedure known as 'burn in'. Confirmation of convergence of the Markov chain is achieved by inspection of the trajectories to check that there is no obvious remaining trend, or by running several independent simulations to verify that the various solutions lie within a reasonable range.

One example of the NMR reconstruction problem employs the *reversible-jump* Markov chain Monte-Carlo method [16]. It assumes that the model spectrum $S(F_1, F_2)$ is made up of a limited number m of two-dimensional Gaussian resonance lines. Then m, the linewidths, intensities, and frequency co-ordinates are varied until the Markov chain reaches convergence. The allowed transitions between the current map M and the new map M' comprise movement, merging or splitting of resonance lines, and 'birth' or 'death' of component responses. Compatibility with the experimental traces is checked by projecting M' at the appropriate angles. The procedure has been found to be stable and reproducible [16].

Some measure of the reliability of all these statistical methods can be obtained by rerunning the programs with different initial conditions. It emerges that in general the location of peaks in the reconstruction is well reproduced, but relative intensities can sometimes vary appreciably. The possibility of false or missing correlations suggests that, in principle, the aforementioned deterministic schemes may be preferable.

4 Four-Dimensional Spectroscopy

When there is ambiguity in the three-dimensional spectrum, or where global isotopic enrichment in ^{13}C and ^{15}N has been employed, a further evolution dimension may be introduced [18]. The problem can still be visualized as a cube in three-dimensional *evolution space*, neglecting any representation of the real-time direct acquisition dimension t_4. The three evolution parameters are defined by

$$t_1 = t \cos\alpha \cos\beta, \tag{13}$$

$$t_2 = t \sin\alpha \cos\beta, \tag{14}$$

$$t_3 = t \sin\beta. \tag{15}$$

(These reduce to the expressions for three-dimensional spectra if $\beta = 0°$.) After Fourier transformation a cube that represents the evolution subspace $S(F_1F_2F_3)$ is created, with the fourth frequency dimension F_4 left to the imagination. In this representation the simplest projections are the three 'first planes' F_1F_4 (where $t_2 = t_3 = 0$), F_2F_4 (where $t_1 = t_3 = 0$), and F_3F_4 (where $t_1 = t_2 = 0$). Resonance locations in one such plane are independent of peak positions in one of the other planes. Normally these first planes do not provide enough information to solve the reconstruction problem unambiguously. However they do generate accurate values of the chemical shifts, they tend to have relatively good sensitivity, and they can be 'borrowed' from related NMR experiments if necessary. A second category of projections is generated by varying two evolution parameters (say t_1 and t_2) in step, while holding the third (t_3) at zero. There are three such kinds of tilted projections, at angles $\pm\alpha$ with $\beta = 0°$, at $\pm\beta$ with $\alpha = 0°$, and at $\pm\beta$ with $\alpha = 90°$.

The third category comprises doubly-tilted projections (involving simultaneous tilting through α and β) recorded when t_1, t_2, and t_3 are incremented jointly. The observed NMR signals are modulated as functions of the evolving chemical shifts (Ω_A, Ω_B, and Ω_C). There are now eight relevant time-domain expressions:

$$S_1 = \cos(\Omega_A t \cos\alpha\cos\beta)\,\cos(\Omega_B t \sin\alpha\cos\beta)\,\cos(\Omega_C t \sin\beta), \tag{16}$$

$$S_2 = \cos(\Omega_A t \cos\alpha\cos\beta)\,\cos(\Omega_B t \sin\alpha\cos\beta)\,\sin(\Omega_C t \sin\beta), \tag{17}$$

$$S_3 = \cos(\Omega_A t \cos\alpha\cos\beta)\,\sin(\Omega_B t \sin\alpha\cos\beta)\,\cos(\Omega_C t \sin\beta), \tag{18}$$

$$S_4 = \cos(\Omega_A t \cos\alpha\cos\beta)\,\sin(\Omega_B t \sin\alpha\cos\beta)\,\sin(\Omega_C t \sin\beta), \tag{19}$$

$$S_5 = \sin(\Omega_A t \cos\alpha\cos\beta)\,\cos(\Omega_B t \sin\alpha\cos\beta)\,\cos(\Omega_C t \sin\beta), \tag{20}$$

$$S_6 = \sin(\Omega_A t \cos\alpha\cos\beta)\,\cos(\Omega_B t \sin\alpha\cos\beta)\,\sin(\Omega_C t \sin\beta), \tag{21}$$

$$S_7 = \sin(\Omega_A t \cos\alpha\cos\beta)\,\sin(\Omega_B t \sin\alpha\cos\beta)\,\cos(\Omega_C t \sin\beta), \tag{22}$$

$$S_8 = \sin(\Omega_A t \cos\alpha\cos\beta)\,\sin(\Omega_B t \sin\alpha\cos\beta)\,\sin(\Omega_C t \sin\beta). \tag{23}$$

The treatment is a straightforward extension of the three-dimensional case outlined in Sect. 2. There are four independent projections of the four-dimensional spectrum $S(F_1F_2F_3F_4)$, giving the frequencies of a typical peak symmetrically related in two pairs:

$$\Omega_A \cos\alpha\cos\beta + \Omega_B t \sin\alpha\cos\beta + \Omega_C t \sin\beta, \tag{24}$$

$$\Omega_A \cos\alpha\cos\beta - \Omega_B t \sin\alpha\cos\beta + \Omega_C t \sin\beta, \tag{25}$$

$$\Omega_A \cos\alpha\cos\beta + \Omega_B t \sin\alpha\cos\beta - \Omega_C t \sin\beta, \tag{26}$$

$$\Omega_A \cos\alpha\cos\beta - \Omega_B t \sin\alpha\cos\beta - \Omega_C t \sin\beta. \tag{27}$$

5 Hyperdimensional Spectroscopy

The treatment in Sect. 4 is readily extended to five dimensions [18, 34], but the time factor begins to be critical for actual measurements. Drastic economies in digitization must be made in all four evolution intervals before the experiment becomes practically feasible. A five-dimensional experiment that employs *only* 16 complex time-domain samples in each of the four evolution periods, with 1 s allowed for (complex) signal acquisition and relaxation, would require 12 days to complete. Furthermore, there would be cumulative losses of magnetization due to relaxation and pulse imperfections, and a fourfold overall signal loss attributable to the $\sqrt{2}$ attenuation between successive stages. Even the processing and storage of high-dimensional data begins to make excessive demands on present-day computers. Although such an experiment is feasible in practice, it is far better to consider an alternative mode for higher dimensional spectra.

The new concept is called *hyperdimensional NMR* [35, 36]. Consider the case of a ten-dimensional experiment, as might be contemplated for a ten-spin system representing two adjacent aminoacids in a large biomolecule. Imagine the corresponding *virtual* matrix comprising all ten orthogonal frequency dimensions. There is no point in attempting to construct this matrix by means of an *actual* ten-dimensional experiment, but it can be used as a conceptual framework for combining lower-dimensional results. The key point is that (say) a three-dimensional spectrum and a four-dimensional spectrum can be combined into a six-dimensional spectrum *provided* they share one common frequency axis. (A minor assignment problem arises if there are degenerate chemical shifts in the common dimension, but there are relatively simple solutions to this difficulty [37]). Tacking together the appropriate low-dimensional spectra on this imaginary framework allows any one of the ten chemical sites to be correlated with any other; there are 45 pairwise correlations of this kind. Note the irony that the results for a ten-dimensional problem are only easily visualized as *plane projections* of this monster matrix. The conventional procedure has always been manual cross-referencing of peaks in the two independent low-dimensional spectra. In contrast, hyperdimensional NMR combines these spectra directly, and then relates them to a virtual high-dimensional matrix.

As a practical illustration, Fig. 5 shows 4 typical two-dimensional projections of the ten-dimensional spectrum of a small 39-residue protein agitoxin, globally enriched in ^{13}C and ^{15}N. All these spectra were obtained by combining three- and four-dimensional experiments that were completed in a reasonably short time, whereas the duration of the full ten-dimensional experiment would have been completely unacceptable. These four planes have been selected from the full complement of 45 possible projections. Each of these spectra contains many cross-peaks because there are many different pairs of adjacent amino acids.

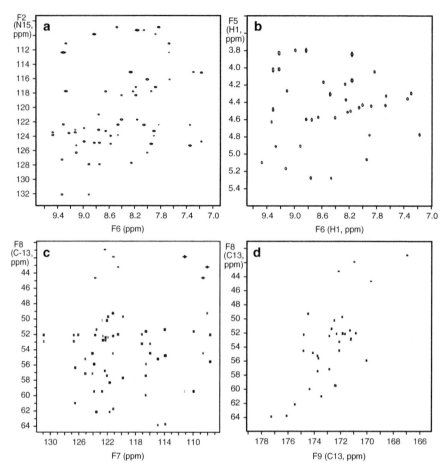

Fig. 5 Four typical planes chosen from 45 possible projections of a virtual ten-dimensional matrix, representing the ten-dimensional spin systems in adjacent aminoacid residues of a small protein, agitoxin. They show the correlations (**a**) N($i-1$) to NH(i), (**b**) CH($i-1$) to NH(i), (**c**) N(i) to Cα(i), (**d**) Cα(i) to CO(i)

6 Conclusions

Projection-reconstruction is not a new idea. The brain performs hundreds of related operations every second by constructing mental three-dimensional images based on two slightly shifted two-dimensional views of the outside world. Applications in other scientific fields – magnetic resonance imaging and X-ray tomography – are well known. This review focuses on the data-sampling methods required to implement projection-reconstruction schemes designed to speed up multidimensional

NMR spectroscopy. Radial sampling of time-domain data is clearly an effective sparse sampling route for this purpose. It relies on a well-proven theorem that the Fourier transform of a skew slice through a two-dimensional time-domain function is the projection of the corresponding frequency-domain function viewed at the same angle. Although, as with all sparse sampling protocols, this scheme introduces artefacts, these are well defined and can be suppressed very effectively. There are basically four deterministic schemes to implement the reconstruction stage. It is essential to match the mode of reconstruction to the appropriate experimental situation – the additive scheme for sensitivity, the lowest-value program for artefact suppression, the Mandelshtam hybrid algorithm for balancing the accumulation and purging features, or the algebraic algorithm for complicated biochemical spectra. On the whole the deterministic schemes are to be preferred over statistical model-fitting procedures. Finally the review describes an effective way to deal with very high dimensional cases – hyperdimensional spectroscopy. Note that the term is not merely a codeword for experiments in higher dimensions, but a conceptual framework for dealing with such systems and for extracting the appropriate information.

References

1. Hernández C, Vogiatzis G, Cipolla R (2008) Multi-view photometric stereo. IEEE Trans Pattern Anal Mach Intell 30:548–554
2. Hounsfield GN (1973) Brit J Radiol 46:1016
3. Lauterbur PC (1973) Nature (London) 242:190
4. Freeman R, Kupče E (2003) J Biomol NMR 27:101–113
5. Bracewell RN (1956) Austr J Phys 9:198
6. Nagayama K, Bachmann P, Wüthrich K, Ernst RR (1978) J Magn Reson 31:133–148
7. Bodenhausen G, Ernst RR (1981) J Magn Reson 45:367–373
8. Bodenhausen G, Ernst RR (1982) J Am Chem Soc 104:1304–1309
9. Ding K, Gronenborn A (2002) J Magn Reson 156:262–268
10. Kim S, Szyperski T (2003) J Am Chem Soc 125:1385–1393
11. Kozminski W, Zhukov I (2003) J Biomol NMR 26:157–166
12. Kupče E, Freeman R (2003) J Biomol NMR 27:383–387
13. Kupče E, Freeman R (2003) J Am Chem Soc 125:13958–13959
14. Deans SR (1983) The Radon transform and some of its applications. Wiley, New York
15. Kupče E, Freeman R (2004) Concepts Magn Reson 22A:4–11
16. Yoon JW, Godsill S, Kupče E, Freeman R (2006) Magn Reson Chem 44:197–209
17. Kupče E, Freeman R (2004) J Biomol NMR 28:391–395
18. Kupče E, Freeman R (2004) J Am Chem Soc 126:6429–6440
19. Coggins BE, Venters RA, Zhou P (2005) J Am Chem Soc 127:11562
20. Ables JG (1974) Astron Astrophys Suppl 15:383
21. Högbom JA (1974) Astron Astrophys Suppl 15:417
22. Shaka AJ, Keeler J, Freeman R (1984) J Magn Reson 56:294
23. Kupče E, Freeman R (2005) J Magn Reson 173:317–321
24. Baumann R, Wider G, Ernst RR, Wüthrich K (1981) J Magn Reson 44:402
25. McIntyre L, Wu X-L, Freeman R (1990) J Magn Reson 87:194–201
26. Venters RA, Coggins BE, Kojetin D, Cavanagh J, Zhou P (2005) J Am Chem Soc 127:8785
27. Ridge CD, Mandelstam V (2010) J Biomol NMR 43:51–159

28. Liang Z-P, Lauterbur PC (2000) Principles of magnetic resonance imaging. A signal processing perspective. IEEE Press, New York
29. Hiller S, Fiorito F, Wüthrich K, Wider G (2005) Proc Natl Acad Sci USA 102:10876
30. Metropolis N, Rosenbluth AW, Rosenbluth MN, Teller AH (1953) J Chem Phys 21:1087
31. Dempster AP, Laird NM, Rubin DB (1977) J Roy Stat Soc 39:1
32. Andrieu C, De Freitas N, Doucet A, Jordan MI (2003) Mach Learn 50:5
33. Green P (1995) Biometrica 82:711
34. Freeman R, Kupče E (2004) Concepts Magn Reson 23A:63–75
35. Kupče E, Freeman R (2006) J Am Chem Soc 128:6020–6021
36. Kupče E, Freeman R (2008) Progr NMR Spectrosc 52:22–30
37. Kupče E, Freeman R (2008) J Magn Reson 191:164–168

Automated Projection Spectroscopy and Its Applications

Sebastian Hiller and Gerhard Wider

Abstract This chapter presents the NMR technique APSY (automated projection spectroscopy) and its applications for sequence-specific resonance assignments of proteins. The result of an APSY experiment is a list of chemical shift correlations for an N-dimensional NMR spectrum ($N \geq 3$). This list is obtained in a fully automated way by the dedicated algorithm GAPRO (geometric analysis of projections) from a geometric analysis of experimentally recorded, low-dimensional projections. Because the positions of corresponding peaks in multiple projections are correlated, thermal noise and other uncorrelated artifacts are efficiently suppressed. We describe the theoretical background of the APSY method and discuss technical aspects that guide its optimal use. Further, applications of APSY-NMR spectroscopy for fully automated sequence-specific backbone and side chain assignments of proteins are described. We discuss the choice of suitable experiments for this purpose and show several examples. APSY is of particular interest for the assignment of soluble unfolded proteins, which is a time-consuming task by conventional means. With this class of proteins, APSY-NMR experiments with up to seven dimensions have been recorded. Sequence-specific assignments of protein side chains in turn are obtained from a 5D TOCSY-APSY-NMR experiment.

Keywords APSY · Automated peak picking · Automated resonance assignment · GAPRO · NMR · Projection spectroscopy · Protein backbone · Protein side chains

S. Hiller
Biozentrum, Universität Basel, Klingelbergstr. 70, 4056 Basel, Switzerland
e-mail: sebastian.hiller@unibas.ch

G. Wider (✉)
Institut für Molekularbiologie und Biophysik, ETH Zürich, Schafmattstr. 20, 8093 Zürich, Switzerland
e-mail: gsw@mol.biol.ethz.ch

Contents

1 Introduction 22
2 Theoretical Background 23
   2.1 The Projection–Cross-Section Theorem 23
   2.2 Projections of Cross Peaks 25
   2.3 The APSY Procedure 26
   2.4 The Secondary Peak Filter 29
3 Practical Aspects 29
   3.1 Sensitivity for Signal Detection in APSY-NMR Experiments 29
   3.2 Sensitivity and Speed of APSY-NMR Experiments 31
   3.3 Selection of Projection Angles 31
   3.4 Optimizing the GAPRO Parameters for a Given Experiment 33
   3.5 Selection of the Number of Projections 34
4 APSY-Based Automated Resonance Assignments 35
   4.1 Overview 35
   4.2 Combinations of 4D and 5D APSY-NMR Experiments 35
   4.3 Backbone Assignments with a Single 6D APSY-NMR 37
   4.4 7D APSY-NMR Spectroscopy for the Assignment of Non-Globular Proteins . 37
   4.5 Automated NMR Assignment of Protein Side Chain Resonances 41
5 Conclusion and Outlook 43
References 44

Abbreviations

1D (2D, 3D, 4D, 5D, 6D, 7D)	One- (two-, three-, four-, five-, six-, seven-) dimensional
ALASCA	Algorithm for local and linear assignment of side chains from APSY data
APSY	Automated projection spectroscopy
GAPRO	Geometric analysis of projections
NMR	Nuclear magnetic resonance
TOCSY	Total correlation spectroscopy

1 Introduction

In NMR studies of biological macromolecules in solution [1–4], multidimensional NMR data are commonly acquired by sampling the time domain in all dimensions equidistantly [5]. With recent advances in sensitivity, such as high field strengths and cryogenic detection devices, the time required to explore the time domain in the conventional way often exceeds the minimal experiment time required by sensitivity considerations, so that the desired resolution determines the duration of the experiment. This situation of the "sampling limit," is common in three- and higher-dimensional experiments with small and medium-sized proteins [6].

When working in the sampling limit, it is worthwhile to obtain the spectral information by "unconventional" experimental schemes, such as non-uniform sampling of the time domain [7–9] or by combination of two or more indirect dimensions [10–12]. The latter approach is also the basis for projection-reconstruction (PR-) NMR [13–16], where the projection–cross-section theorem [17, 18] is combined with image reconstruction techniques [19, 20] to reconstruct the multidimensional frequency domain spectrum from experimentally recorded projections. Further practical acquisition and processing techniques for unconventional multidimensional NMR experiments have been demonstrated [21–33]. Several of these methods are discussed in the other chapters of this book.

The analysis of NMR spectra involves intensive human intervention, and automation of NMR spectroscopy with macromolecules is thus of general interest. Major challenges are the distinction of real resonance peaks from thermal noise and spectral artifacts, as well as peak overlap [34–36]. On grounds of principle, automated analysis benefits from higher-dimensionality of the spectra [21, 37], since the peaks are then more widely separated, and hence peak overlap is substantially reduced.

APSY (automated projection spectroscopy) combines the technique to record projections of high-dimensional NMR experiments [15] with automated peak-picking of the projections and a subsequent geometric analysis of the peak lists with the algorithm GAPRO (geometric analysis of projections). Based on geometrical considerations, GAPRO identifies peaks in the projections that arise from the same resonance in the N-dimensional frequency space, and subsequently calculates the positions of these peaks in the N-dimensional spectral space. The output of an APSY-experiment is thus an N-dimensional chemical shift correlation list of high quality which allows efficient and reliable subsequent use by computer algorithms. Due to extensive redundancy in the input data for GAPRO, high precision of the chemical shift measurements is achieved. Importantly, APSY is fully automated and operates without the need to reconstruct the high-dimensional spectrum at any point.

In the following sections, the theoretical and practical foundations of APSY are introduced. Several practical aspects are discussed including the sensitivity of APSY experiments. Then, applications of APSY for the assignment of protein resonances are described. For the backbone assignment, the high-quality APSY peak lists are used as the input for a suitable automatic assignment algorithm. For example, the 6D APSY-seq-HNCOCANH experiment connects two sequentially neighboring amide moieties in polypeptide chains via the ^{13}C' and $^{13}C^\alpha$ atoms. Further applications are the backbone assignment of unfolded proteins and the side chain assignment of folded proteins.

2 Theoretical Background

2.1 The Projection–Cross-Section Theorem

The projection–cross-section theorem states that an m-dimensional cross section, $c_m(t)$, through N-dimensional time domain data ($m < N$) is related by an m-dimensional Fourier transformation to an m-dimensional orthogonal projection of the N-dimensional NMR spectrum, $P_m(\omega)$, in the frequency domain [17, 18]. Thereby, $P_m(\omega)$ and $c_m(t)$ are oriented by the same angles with regard to their corresponding coordinate systems (Fig. 1).

Kupče and Freeman showed that this theorem can be utilized to record projections of multidimensional NMR experiments. The time domain is sampled along a straight line (Fig. 1) and quadrature detection for this cross-section $c_m(t)$ is obtained by combing data from corresponding positive and negative projection angles using the trigonometric addition theorem [14, 15]. The subsequent hypercomplex Fourier transformation results in the projections $P_m(\omega)$ [11, 15, 22]. Projections with a dimensionality of $m = 2$, with one directly recorded and one indirect dimension, are the most practical case. For such 2D projections, the indirect dimension is a 1D projection of the $N - 1$ indirect dimensions of the N-dimensional experiment. The orientations of both $c_2(t)$ and $P_2(\omega)$ are described by $N - 2$ projections angles.

For example, in a 5D APSY experiment ($N = 5$), three projection angles α, β and γ, define the orientations of $c_2(t)$ and $P_2(\omega)$. The two unit vectors $\vec{p}_1$ and $\vec{p}_2$, which span the indirect and the direct dimension, respectively, are given by

$$\vec{p}_1 = \begin{pmatrix} \sin \gamma \\ \sin \beta \cdot \cos \gamma \\ \sin \alpha \cdot \cos \beta \cdot \cos \gamma \\ \cos \alpha \cdot \cos \beta \cdot \cos \gamma \\ 0 \end{pmatrix} \quad \vec{p}_2 = \begin{pmatrix} 0 \\ 0 \\ 0 \\ 0 \\ 1 \end{pmatrix}. \quad (1)$$

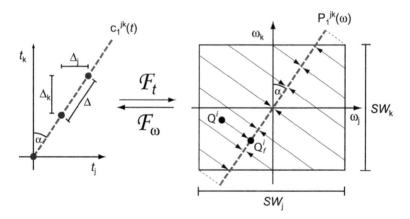

Fig. 1 Illustration of the projection–cross-section theorem [17–19] for a 2D frequency space with two indirect dimensions k and j. 1D data $c_1^{jk}(t)$ on a straight line in the 2D time domain (t_j, t_k) (*left*) is related to a 1D orthogonal projection $P_1^{xy}(\omega)$ of the spectrum in the 2D frequency domain (ω_j, ω_k) (*right*) by a 1D Fourier transformation, F_t, and the inverse transformation, F_ω. The projection angle α describing the slope of $c_1^{jk}(t)$ defines also the slope of $P_1^{xy}(\omega)$. The cross peak Q^i (*black dot*) appears at the position Q_f^i in the projection. Further indicated are the spectral widths in the two dimensions of the frequency domain, SW_j and SW_k, and the evolution time increments Δ, Δ_k and Δ_j (1)–(4). Adapted with permission from [38]

An N-dimensional NMR spectrum is spanned by user-defined sweep widths SW_i for each of the N dimensions ($i = 1,\ldots, N$). For a projection spectrum defined by $\vec{p}_1$, an appropriate sweep width SW needs to be calculated (Fig. 1). Considering that the distribution of chemical shifts in a given dimension is well described by a normal distribution [39], the sweep width can be calculated as [40]

$$SW = \sqrt{\sum_{i}^{N-1}(SW_i \cdot p_i)^2}, \quad (2)$$

where p_1^i are the coordinates of the vector $\vec{p}_1$ (1). The dwell time for the recording of discrete data points, Δ, is then calculated as

$$\Delta = 1/SW, \quad (3)$$

and the resulting increments for the $N - 1$ evolution times t_i, Δ_i, in the $N - 1$ indirect dimensions (Fig. 1), are given by

$$\Delta_i = p_1^i \cdot \Delta. \quad (4)$$

For an optimal phasing of the projection spectra in the N-dimensional space it is advisable to sample the time domain starting at the origin (Fig. 1). Thus, all APSY pulse sequences should allow sampling access to this time domain point with zero evolution time.

2.2 Projections of Cross Peaks

In a set of j projections with different projection vectors $\vec{p}_{1,f}$, an N-dimensional cross peak Q^i is projected orthogonally to the locations Q_f^i. Here, f is an arbitrary numeration of the set of j projections $f = 1, \ldots, j$. In the 2D coordinate system of projection f, the projected cross peak has the position vector $\vec{Q}_f^i = [v_{f,1}^i, v_{f,2}^i]$, with $v_{f,1}^i$ and $v_{f,2}^i$ being the chemical shifts along the projected indirect dimension and the direct dimension, respectively. It is convenient to define the origins of both the N-dimensional coordinate system and the 2D coordinate system in all dimensions in the center of the spectral ranges. Then the position vector $\vec{Q}_f^i$ in the N-dimensional frequency space is given by

$$\vec{Q}_f^i = v_{f,1}^i \cdot \vec{p}_{1,f} + v_{f,2}^i \cdot \vec{p}_2. \quad (5)$$

The N-dimensional cross peak Q^i is located in an $(N - 2)$-dimensional subspace, which is orthogonal to the projection plane at the point Q_f^i (Fig. 1). The "peak subgroup" of an N-dimensional chemical shift correlation Q^i is the set of projected peaks, $\{Q_1^i, \ldots, Q_j^i\}$, that arise from it. It is the key function of the GAPRO algorithm to identify the peak subgroups in the j peak lists of the projections and to calculate the coordinates of Q^i from them.

2.3 The APSY Procedure

The APSY procedure follows the flow-chart shown in Fig. 2. It is illustrated in Fig. 3 and the APSY input parameters are defined in Table 1. At the start, the operator selects the desired N-dimensional NMR experiment, the dimensionality of the projection spectra, and j sets of projection angles. The projection spectra are recorded and automatically peak picked using the GAPRO peak picker, resulting in j peak lists. The GAPRO peak picker identifies all local maxima of the spectrum with a sensitivity (signal-to-noise) larger than a user-defined value R_{min}. The position of the maximum is interpolated for each peak by a symmetrization procedure that involves the intensities of the two neighboring data points in each dimension [40]. The GAPRO peak picker does not attempt to distinguish real peaks from spectral artifacts or random noise; every local maximum is identified as a peak. The j peak lists contain peaks Q_{gf}, where g is an arbitrary numeration of the peaks and f of the projections ($f = 1, \ldots, j$). GAPRO then arbitrarily selects $N - 1$ of these peak lists, and generates for each peak Q_{gf} a subspace L_{gf}, which contains the point Q_{gf} and which is orthogonal to the projection f (Fig. 3b). The intersections of the subspaces L_{gf} in the N-dimensional space are candidates for the positions of N-dimensional cross peaks (open circles in Fig. 3b). To account for the imprecision in the picked peak positions due to thermal noise, the calculation of intersections of

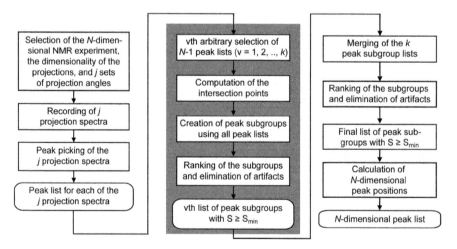

Fig. 2 Flowchart of the APSY procedure. *Square boxes* indicate processes, and *boxes with rounded corners* denote intermediate or final results. The steps *surrounded with gray* are repeated k times, and thus generate k lists of peak subgroups. Adapted with permission from [38]

Automated Projection Spectroscopy and Its Applications

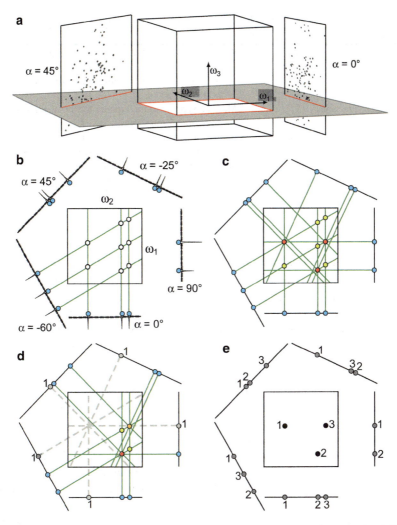

Fig. 3 Illustration of the algorithm GAPRO for $N = 3$, $j = 5$, $k = 1$ and $S_{min} = 3$. (**a**) Three dimensional view of the situation with the unknown 3D spectrum shown as a *cube* in the center and two 2D projections with $\alpha = 0°$ and $\alpha = 45°$. (**b–e**) These panels are oriented like the *gray* ω_1/ω_2-plane in (**a**). (**b**) A 2D cross section through the unknown 3D spectrum is surrounded by 1D cross sections through the five experimental 2D projections with projection angles $\alpha = 0°$, $90°$, $-30°$, $45°$, and $-65°$. The *cyan dots* mark the result of the automatic peak picking of the 2D projections. The algorithm then arbitrarily selected $N - 1 = 2$ of the j projections for the first round of spectral analysis, with $\alpha = 0°$ and $\alpha = -60°$. The intersections of the subspaces corresponding to the peaks in these two projections (*green lines*) identify eight candidate points in the 3D spectrum (*open circles*). (**c**) Using the subspaces from all five projections, the support S (number of intersecting subspaces, see text) is calculated for each candidate point. *Yellow* and *red dots* indicate $S = 2$ and $S = 5$, respectively. (**d**) One of the three candidate points with the highest support ($S = 5$) is arbitrarily selected. All peaks in the projections that contribute to the selected candidate point are identified as a peak subgroup (*gray dots* in the projections labeled with number 1). The subspaces from this subgroup are removed from the further analysis (*gray dashed lines*).

Table 1 Parameters used as part of the input for the software GAPRO

Subroutines	Parameters	Description
Peak picking	R_{min}	Signal-to-noise threshold for peak picking
	$\Delta\omega_{H2O}$	Frequency range along the acquisition dimension on both sides of the water resonance within which no peaks are picked
GAPRO core functions	S_{min}	Minimal support value required for the identification of an N-dimensional peak
	Δv_{min}	Intersection tolerance in the directly detected dimension
	r_{min}	Intersection tolerance in the indirect dimensions
	k	Number of independent GAPRO calculations from which the final result is derived
Secondary peak filter	R_{single}	Signal-to-noise threshold at the back-projected peak positions (see text)
	n	Allowed number of violations of R_{single}

subspaces allows a user-defined tolerance value in the direct dimension, Δv_{min}. For each of the candidate points, the support, S, is then calculated. S is the number of subspaces from all j projections that contain the candidate point. Thereby at most one subspace from each projection is considered (Fig. 3c) so that $N - 1 \leq S \leq j$. For the calculation of the support, in addition to the user-defined tolerance values for the direct dimension, Δv_{min}, a tolerance for the indirect dimension, r_{min}, is also required to account for the imprecision of picked peak positions. The geometric analysis algorithm can also include aliased peaks in the experimental 2D projections at this point of the calculation. The peaks that contribute to the support of a given candidate point form a "peak subgroup." The subgroups are ranked for high S-values, and the top-ranked subgroup is selected. In case of degeneracy, one of the top-ranked subgroups is arbitrarily selected. The subspaces contributing to this subgroup are removed from further analysis, and new S-values for the residual candidate points are calculated from the remaining subspaces (Fig. 3c). This procedure is repeated until the value of S for all remaining subgroups falls below a user-defined threshold, S_{min}. At this point in the algorithm a list of the identified peak subgroups is generated. The subgroup identification is repeated with k different, randomly chosen starting combinations of $N - 1$ projections, and k peak subgroup lists are thus obtained (gray box in Fig. 2). These k lists are merged into a single list, which is again subjected to ranking and elimination of all subgroups with $S < S_{min}$. From the resulting final list of subgroups, the peak positions in the N-dimensional space are calculated (Fig. 3d). Since the peak positions are redundantly determined by the experimental data, particularly high precision can

---

Fig. 3 (continued) The support S of remaining candidate points is recalculated (there remains one point with $S = 5$, and another one with $S = 4$ is shown in *orange*). (**e**) After two more rounds of the procedure indicated in (**d**), two additional subgroups are identified and labeled with numbers 2 and 3, respectively. From the three subgroups, the positions of three peaks in the 3D spectrum are calculated (*black dots*). Adapted with permission from [38]

2.4 The Secondary Peak Filter

In APSY, the discrimination between artifacts and noise corresponds to distinguishing between peaks in the projections that stem from an N-dimensional resonance and are thus correlated and those that are uncorrelated. With a sufficiently large number of independently recorded experimental projection spectra only true N-dimensional chemical shift correlations are contained in the final peak list. This separation can be further enhanced by applying an additional, secondary peak filter to the final result of the GAPRO calculation. Thereby the N-dimensional APSY peak list is "back-projected" onto the experimental projections, and the spectral sensitivities at the resulting positions are read out. Based on user-defined criteria, the N-dimensional GAPRO peak list can then be filtered to remove weakly supported peaks or remaining artifacts. All peaks with more than n violations of the threshold R_{single} (Table 1) are deleted. The secondary filter thus provides an efficient additional validation of the GAPRO result and permits the use of less stringent parameters in the GAPRO run.

3 Practical Aspects

3.1 Sensitivity for Signal Detection in APSY-NMR Experiments

The intensity of a given multi-dimensional NMR signal varies in the individual projections of an APSY experiment. In the following the expressions for the signal-to-noise ratio of a given resonance are presented. This formalism can then be used to optimize the performance of APSY-NMR experiments.

By adapting general equations for 2D spectra [5] to APSY-NMR [40], the sensitivity of a signal in a projection of an experiment m with projection angles $\vec{\varphi} = (\alpha, \beta, \ldots)$ is given by

$$[S/\sigma]_m(\vec{\varphi}) = K_A \cdot s_m(0) \cdot f_m(\vec{\varphi}), \qquad (6)$$

where the three terms K_A, $s_m(0)$ and $f_m(\vec{\varphi})$ represent, respectively, the impact of the detected spin type A, the signal intensity at time zero, and the dependence on the projection angles $\vec{\varphi}$.

K_A accounts for the properties of the detected nuclear species A (often protons), including the probe sensitivity, the main polarizing magnetic field strength, and the window function applied before Fourier transformation. Thus, the value of K_A can be maximized for a detected given nucleus type A and a given NMR instrument for all experiments that are detected on this nucleus.

$s_m(0)$ is the signal intensity of the experiment m at the time domain origin. This factor enables a comparison of the relative intrinsic sensitivities of different APSY-NMR experiments that are detected on the same nucleus, and can thus help to identify high-sensitivity experiments. Values for $s_m(0)$ can be estimated either experimentally, e.g., from 1D NMR spectra of the time domain origins, or from model calculations [41]. Table 2 lists calculated values for different amide proton-detected experiments.

Finally, $f_m(\vec{\varphi})$ describes the dependence of the sensitivity of an experiment m on the projection angles $\vec{\varphi}$ and on the acquisition and processing parameters [40]:

$$f_m(\vec{\varphi}) = \frac{1}{(\sqrt{2})^q} \frac{\sqrt{n(\vec{\varphi}) \cdot M(\vec{\varphi})}}{\sqrt{h^2(\vec{\varphi})}} \cdot \frac{1}{t_{max}(\vec{\varphi})} \int_0^{t_{max}(\vec{\varphi})} dt \, s_m^e(\vec{\varphi}, t) \cdot h(\vec{\varphi}, t). \quad (7)$$

Here, $s_m^e(\vec{\varphi}, t)$ is the signal envelope in the indirect dimension, q is the number of angles that differ from 0° or 90° (the number of subspectra to be combined for quadrature detection is $2q$ [11, 15]), and $n(\vec{\varphi})$ is the operator-chosen number of scans recorded for each of the subspectra. $h_m(\vec{\varphi}, t)$ is the applied window function, $t_{max}(\vec{\varphi})$ the maximal evolution time, and $M(\vec{\varphi})$ the number of indirect points sampled.

$f_m(\vec{\varphi})$ is largely governed by the envelope function $s_m^e(\vec{\varphi}, t)$. It can be shown that monoexponential relaxation in all indirect dimensions results in $s_m^e(\vec{\varphi}, t)$ being a monoexponential decay with a decay rate constant $R_{2,m}^*(\vec{\varphi})$ given by

$$R_{2,m}^*(\vec{\varphi}) = \vec{p}_1(\vec{\varphi}) \cdot \vec{R}_{2,m}. \quad (8)$$

Table 2 Theoretical sensitivities, $s_m(0)$, of APSY-NMR experiments for polypeptide backbone assignments

Correlated atoms[a]	Experiment	$s_m(0)$[b]
HN–N	2D [^{15}N,^{1}H]-HSQC	100[c]
HN–N–C$^\alpha$–C$^\beta$ seq	5D APSY-HNCOCACB	11
	5D APSY-CBCACONH	10–4[d]
HN–N–C$^\alpha$–C$^\beta$ intra	4D APSY-HNCACB	7
	4D APSY-CBCANH	5–2[d]
HN–N–C$^\alpha$–H$^\alpha$ seq	5D APSY-HNCOCAHA	17
	5D APSY-HACACONH	15
HN–N–C$^\alpha$–H$^\alpha$ intra	4D APSY-HNCAHA	11
	4D APSY-HACANH	8
HN–N–C$^\alpha$–C' seq	4D APSY-HNCOCA	32
HN–N–C$^\alpha$–C' intra	4D APSY-HNCACO	7

[a] seq and intra stand for sequential connectivity and intraresidual connectivity, respectively. [b] The data were calculated for a protein with a rotational correlation time of 10 ns [40]. [c] The sensitivity of the 2D [^{15}N,^{1}H]-HSQC spectrum was normalized to 100. [d] For these experiments, the intensity depends on the amino acid type. The ranges give the minimal and maximal value among the 20 common amino acid residues

Here, $\vec{R}_{2,m}$ is an N-dimensional vector containing the transverse relaxation rates along all indirect dimensions, with $R_{2,m}^i = 0$ for constant-time evolution elements in the dimension i. Since the standard GAPRO analysis attaches equal weight to each projection spectrum, it is desirable to have similar sensitivities for all individual projection experiments. If the projection angle-dependence of $s_m^e(\vec{\varphi}, t)$ is known, (7) provides a basis for producing similar sensitivities for all the projections used in a given APSY experiment, since the user-defined parameters $n(\vec{\varphi})$, $M(\vec{\varphi})$, $h_m(\vec{\varphi}, t)$, and $t_{\max}(\vec{\varphi})$ can be individually adjusted for each projection experiment [5, 42].

3.2 Sensitivity and Speed of APSY-NMR Experiments

With a practical example we want to illustrate the performance of APSY in terms of sensitivity and speed. The example is a 4D APSY-HNCOCA experiment with the 12-kDa protein TM1290, of which the sequence-specific resonance assignments are known [43]. A total of 13 2D projections were measured in 13 min (1 min per projection). The 4D APSY-HNCOCA experiment was recorded with $[U\text{-}^{13}\text{C},^{15}\text{N}]$-labeled TM1290 at 25 °C on a 600 MHz Bruker Avance III spectrometer with a room temperature probe. The concentration was adjusted to 1.0 ± 0.05 mM, as determined by PULCON [44]. The 13 pairs of projection angles (α, β) comprised: $(90°, 0°), (0°, 0°), (0°, 90°), (\pm 60°, 0°), (0°, \pm 60°), (90°, \pm 60°)$, and $(\pm 20°, \pm 70°)$.

In the 4D peak list generated by the algorithm GAPRO from these 13 projections, all 110 expected 4D $(\omega_1(^{15}\text{N}), \omega_2(^{13}\text{C}'), \omega_3(^{13}\text{C}^\alpha), \omega_4(^{1}\text{H}^{\text{N}}))$ chemical shift correlations were contained. With the selected short measuring time, the intensity of the weakest NMR signals in the projections is comparable to the intensity of the thermal noise (Fig. 4). Nonetheless, even the weakest of the 110 correlation peaks (indicated by arrows in Fig. 4) was recognized by GAPRO as a true correlation, whereas no false 4D correlations appeared. This shows that APSY makes use of the combined sensitivity of all the projections in the input, and that it does not require unambiguous identification of the individual peaks in each projection.

3.3 Selection of Projection Angles

The APSY method does not impose restrictions on the choices of the projection angles or the number of projections, except that, on fundamental grounds, the total number of projections must be at least $N - 1$ and that each indirect dimension needs to be evolved at least once in the set of projections. On this basis, the selection of projection angles for a given APSY-NMR experiment should be guided by two main considerations. First, the projection angles should be distributed about evenly in the time domain. Second, projections with large q values (number of projection angles that differ from $0°$ or $90°$) are to be disfavored, since the

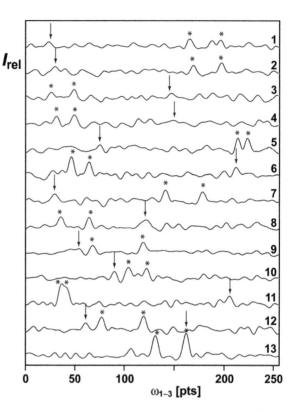

Fig. 4 One-dimensional cross-sections through 13 2D projections of a 4D APSY-HNCOCA experiment of the protein TM1290. The data set was recorded in a total measuring time of 13 min on a Bruker Avance III 600 MHz spectrometer equipped with a room temperature probe head, with a 1 mM protein concentration at 25 °C. The cross sections were taken through the ω_4-position the weakest of the 110 peaks of TM1290. In each of the 13 cross-sections, the position of this weakest peak is indicated with an *arrow*. Asterisks denote two other resonances that are present at this ω_4-position. All other local maxima seen in the cross-sections arise from random spectral noise. Reproduced with permission from [40]

sensitivity for the recording of the 2D projection spectra is proportional to $2^{-q/2}$ (7). It is further recommended that the decrease in sensitivity due to higher q-values is compensated by adjusting the number of scans $n(\vec{\varphi})$ accordingly (7).

An additional improvement is achieved with the use of dispersion-optimized projection angles, in particular if the sweep widths of the indirect dimensions are significantly different. Dispersion-optimized projection angles adjust the contributions of the indirect dimensions to the same size, and thus contribute to eliminating chemical shift overlap. The dispersion-optimized, or "matching" projection angle α^* for two dimensions, i and j, with sweep widths SW_i and SW_j is given by

$$\tan \alpha^* = \frac{SW_i}{SW_j}. \qquad (9)$$

For example, if the sweep widths of two dimensions differ 11-fold (as they do for C' and C^β), then $\alpha^* = 84°$. A set of three projection angles with values of 60°, 84°, and 87° would thus be a good choice, whereas a seemingly more symmetric selection with angles of 22.5°, 45°, and 67.5° would lead to two basically identical projections [40]. Expressions similar to (9) can be derived for combinations of three or more dimensions.

3.4 Optimizing the GAPRO Parameters for a Given Experiment

Among the input variables of the geometric algorithm GAPRO, three parameters have a dominant effect on the result of the spectral analysis: S_{min}, Δv_{min}, and r_{min} (Table 1). The selection of the minimal support S_{min} is most important, since only candidate signals with a support $S \geq S_{min}$ will be included in the final peak list. Figure 5a shows the variation of the peak list resulting from a 4D APSY-HNCOCA experiment when different values of S_{min} are used. The data set consisted of the 13 projections recorded with the protein TM1290 mentioned above, for which 110 amino acids are expected. For S_{min} between 3 and 8, the final result contains all the expected peaks. As a general guideline, it is advisable to set S_{min} to about one third of the number of input projections, and to keep $S_{min} > (N + 2)$ for an N-dimensional experiment with 2D projections.

The two additional key parameters are the intersection tolerances for the direct and indirect dimensions, Δv_{min} and r_{min}. A variation of these parameters shows that

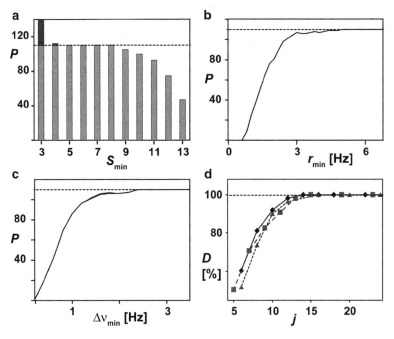

Fig. 5 Impact of APSY parameters on the result. (**a**) Dependence of the total number P of 4D APSY-HNCOCA peaks of the protein TM1290 on the GAPRO parameter S_{min}. *Light gray bars* represent the number of correct correlation peaks, *dark gray bars* the number of artifacts, and the *dotted line* indicates the expected 110 peaks. (**b, c**) Dependence of the total number P of 4D APSY-HNCOCA peaks of the protein TM1290 on the GAPRO parameters r_{min} and Δv_{min}, respectively. The *dotted line* indicates the expected 110 peaks. (**d**) Impact of the number j of 2D projections on the percentage of the expected correlations, D, for three APSY experiments with the protein TM1290. 4D APSY-HACANH (*squares*), 5D APSY-HACACONH (*diamonds*) and 5D APSY-CBCACONH (*triangles*). The data correspond to Table 3. Reproduced with permission from [40]

each of these tolerances has to be larger than a certain minimal value, which depends on the digital resolution, the signal line widths and the sensitivity of a given experiment (Fig. 5b,c). If the tolerances are chosen too small, corresponding subspaces do not intersect and correct peaks are not found. However, if the parameters are too large, no negative effects occur in the final result except that the computation time increases substantially due to the increasing number of intersection possibilities. As a general guideline, it is advisable to use one to two times the respective digital resolutions in the direct and indirect dimensions of the 2D projection spectrum as values for Δv_{min} and r_{min}, respectively.

3.5 Selection of the Number of Projections

Choosing the minimal number of projections needed for a given experiment is important to minimize the required instrument time. A good decision on the number of projections considers the type of APSY-NMR experiment, the expected number of correlation peaks per amino acid residue, the size and type of protein under study, the choice of the projection angles, and the required quality of the result. Representative examples for the number of projections needed in particular experiments are shown in Fig. 5d and Table 3. For polypeptides with smaller chemical shift dispersion, such as denatured proteins, a higher number of projections is required for obtaining comparable results as for globular proteins. It should be noted that APSY can be run using a convergence scheme which interleaves the recording of new projections with the analysis of the existing data by GAPRO. The convergence scheme stops the data recording as soon as a preset number of peaks have been resolved or when the protein has been assigned.

Table 3 APSY-based backbone assignment of the protein TM1290[a]

Parameters	4D APSY-HACANH	5D APSY-HACACONH	5D APSY-CBCACONH
Number of 2D projections j	19	18	20
Recording time per projection	20 min	30 min	25 min
S_{min}	8	8	8
Sequential correlations expected[b]	109	**109**	**110**
Sequential correlations observed[b]	88	**109**	**110**
Intraresidual correlations expected[b]	**109**	0	0
Intraresidual correlations observed[b]	**109**	0	0

[a]The measurements were performed with a 1 mM solution of TM1290 at 25 °C on a 750 MHz NMR spectrometer with room temperature probe. [b]Correlations relevant for the automated assignment are indicated in bold

Note further that in many APSY experiments, some of the projections can be measured with individually optimized, shortened pulse sequences, which omit magnetization transfers that are not required and hence have improved sensitivity [15]. For example, a direct projection of the $\omega(^{15}N)$-dimension in a multidimensional APSY experiment can be replaced by a standard [^{15}N,^{1}H]-HSQC experiment.

4 APSY-Based Automated Resonance Assignments

4.1 Overview

APSY provides peak lists of chemical shift correlations for multidimensional NMR experiments. Due to the averaging of a large number of observed signals in the set of projections, the determination of the N-dimensional chemical shifts becomes very precise. APSY is thus well suited for applications that require precise peak positions. Here we want to concentrate on applications for resonance assignments in protein spectra. APSY-NMR combined with a suitable assignment algorithm enables fully automated sequence-specific assignments for globular and denatured proteins.

4.2 Combinations of 4D and 5D APSY-NMR Experiments

Strategies for sequence-specific backbone resonance assignment of polypeptides usually contain two key elements. First, sequential NMR connectivities lead to the identification of discrete peptide fragments of different lengths. Second, these fragments are mapped onto the known polypeptide sequence, based on the chemical shift statistics of the amino acid types. The vast majority of conventional backbone assignment experiments are detected on the amide proton due to the high experimental sensitivity and other practical aspects. For the same reasons, we also limit the present discussion of APSY experiments to this nucleus. With this selection, the ^{1}H and ^{15}N chemical shifts of the amide moiety are readily contained in each correlation. APSY can connect two sequential amide moieties in a single experiment (see below); however, usually at least two APSY experiments are needed to connect two sequential amide moieties. As illustrated in Fig. 6, the $^{13}C^{\alpha}$ atom is always a nucleus available for the sequential connection, and a second matching nucleus can be either $^{1}H^{\alpha}$, $^{13}C^{\beta}$, or $^{13}C'$. For the mapping of fragments onto the sequence, the $^{13}C^{\beta}$ chemical shift increases the reliability of sequence-specific assignments, since it allows the unambiguous distinction between different amino acid types [45–48]. With the requirement that two chemical shifts should define the sequential connectivities, five groups of four- and five-dimensional correlation experiments can be devised (Fig. 6). The relative sensitivities of the APSY

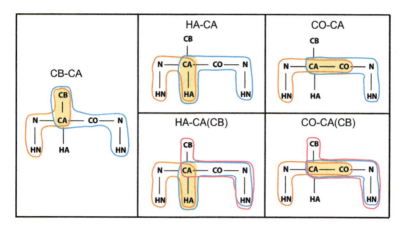

Fig. 6 Combinations of intraresidual and sequential chemical shift correlations to be recorded with H^N-detected APSY-NMR experiments for polypeptide backbone assignments of $^{13}C,^{15}N$-labeled proteins. Each *colored shape* contains the nuclei correlated by a 4D or 5D experiment. The *orange areas* contain the nuclei for which the individual correlations overlap. The notations used for the different groups of experiments are indicated in each panel. Reproduced with permission from [40]

experiments required for these assignment strategies can be estimated by model calculations (Table 2) [40].

As one practical example, we show the application of the HA–CA(CB) strategy to obtain the backbone resonance assignments of the protein TM1290. TM1290 is the same protein as was studied in the experiment of Fig. 5. The HA–CA(CB) strategy is realized with the three experiments 4D APSY-HACANH, 5D APSY-HACACONH, and 5D APSY-CBCACONH. These three experiments were carried out with the same sample of TM1290 as described above and were performed at 25 °C on a 750 MHz Bruker Avance III spectrometer with a room temperature probe. The input for the assignment algorithm GARANT [49] consisted of the three final APSY peak lists and the amino acid sequence [50]. Table 3 presents key parameters used for the recording of these three experiments. The lower part of Table 3 lists the numbers of expected and observed correlations for the three experiments used in the HA–CA(CB) approach. All detected TM1290 backbone resonances were correctly assigned [40]. This example thus shows that in a total instrument time of about 1 day and with minimal human intervention, the complete and correct sequence-specific resonance assignments of a 12-kDa protein were obtained with the APSY-based approach.

For the assignment of larger proteins, the CB–CA and the CO–CA strategies are preferred over the HA–CA strategy, since they are compatible with deuteration, which in turn increases the experimental sensitivity. The CB–CA strategy is realized with the combination of the 4D APSY-HNCACB and the 5D APSY-HNCOCACB experiment [51]. With this approach, the backbones of two human proteins were assigned, the 22-kDa protein kRas at 0.4 mM concentration and a 15-kDa drug target protein (protein A) at 0.3 mM concentration (Fig. 11) [51]. For each of the two proteins, 76 h of experiment time on a 600 MHz spectrometer with a cryogenic probe were used for the 2 backbone experiments (16 projections of the 4D APSY-HNCACB in 48 h and 32 projections of the 5D APSY-HNCOCACB in 48 h) corresponding to a total of 3 days and assigned by the algorithm MATCH. The overall completeness of the backbone resonance assignment was 95% for the 22-kDa protein kRas and 98% for the 15-kDa protein A (Fig. 11), where the missing assignments comprised segments with unfavorable protein dynamics.

4.3 Backbone Assignments with a Single 6D APSY-NMR

The combination of 4D and 5D APSY-NMR experiments thus provides backbone assignments for perdeuterated, [U-^{13}C,^{15}N]-labeled proteins up to at least 25 kDa. For smaller proteins up to 12 kDa an alternative and more elegant approach can be used [50]. The pathway H^N–N–C'–C^α–N–H^N directly connects two sequentially adjacent amide moieties in a single experiment and with APSY the full potential of this magnetization transfer pathway is exploited [50]. The 6D APSY-seq-HNCOCANH experiment was recorded with a 0.9 mM solution of the protein 434-repressor(1–63) at 30 °C on a Bruker 750 MHz spectrometer with room temperature probe. A total of 25 2D projections were recorded in 40 h. The same experiment was also recorded with a 3.0 mM solution of TM1290 at 35 °C on a Bruker DRX 500 MHz spectrometer with a cryogenic probe. The total spectrometer time used for the recording of 25 2D projections was 20 h.

For 434-repressor(1–63), the resulting APSY peak list contained 56 out of 57 peaks expected from the amino acid sequence and for TM1290 all but three of the expected peaks were obtained [43]. Both lists did not contain any artifact and provided very precise chemical shifts. The precision of chemical shift measurements can be directly assessed in this data set, since the resonance of each amide moiety is part of two different 6D peaks. The amide proton chemical shift is measured in the direct dimension ω_6 and in the indirect dimension ω_1; the amide ^{15}N chemical shift in ω_5 and ω_2, respectively. From the 93 amide moieties of TM1290 that contributed to two peaks, the precision (standard deviation) for the proton and nitrogen chemical shift measurements was 0.0014 ppm (0.72 Hz), and 0.0137 ppm (0.69 Hz), respectively. These precise and artifact-free 6D peak lists were used as inputs for the assignment algorithm GARANT [49], yielding the correct sequence-specific assignment for each protein.

4.4 7D APSY-NMR Spectroscopy for the Assignment of Non-Globular Proteins

Studies of soluble non-globular polypeptides are of great relevance for protein folding as well as for insight into the structural basis of functional non-globular polypeptides [52–59]. However, the available data on this class of proteins are scarce because they are not amenable to meaningful single-crystal studies, and solution NMR studies have been limited by small dispersion of the chemical shifts [60–62]. Increased interest in detailed structural and dynamic characterization of soluble non-globular polypeptides has, however, more recently been generated by the discovery of a rapidly increasing number of proteins that are intrinsically unfolded in their functional state in solution [54, 55, 58]. APSY provides a fully automated approach to solving this problem with the use of very high-dimensional NMR.

One approach is a combination of the above-mentioned 6D APSY-seq-HNCOCANH with the 5D APSY-HNCOCACB experiment [63]. Thereby, the 6D APSY scheme [50] connects neighboring amide groups sequentially, and the 5D APSY scheme measures the C^β chemical shifts. These two experiments can also be combined into a single magnetization transfer pathway, the 7D APSY-seq-HNCO(CA)CBCANH (Figs. 7a, and 8) [63]. Magnetization of the amide proton i is transferred with seven subsequent INEPT [64] steps to the amide proton $i - 1$ (steps a–g in Fig. 7a). Along this pathway, six evolution periods are introduced for the frequencies of the nuclei $^1H^N_i$, $^{15}N_i$, $^{13}C'_{i-1}$, $^{13}C^\beta_{i-1}$, $^{13}C^\alpha_{i-1}$, and $^{15}N_{i-1}$. Thus, each seven-atom fragment of residues i and $i - 1$ gives rise to a single peak, except if a proline residue or a chain end is located at either of the positions i or $i - 1$.

The 7D APSY-seq-HNCO(CA)CBCANH is illustrated here with the NMR assignment of the 148-residue outer membrane protein X (OmpX) denatured with 8 M urea in aqueous solution [71]. The experiment was recorded with 100 2D projections in a total measuring time of 2 days (50 h) at 15 °C on a Bruker 750 MHz spectrometer with room temperature probe (Fig. 8) [38]. Out of the 142 expected peaks, 139 were actually observed [63]. The three missing peaks connect

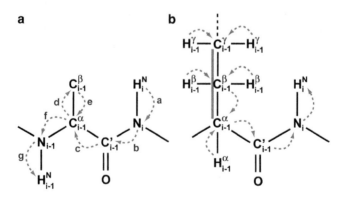

Fig. 7 Magnetization transfer pathways of (a) the 7D APSY-seq-HNCO(CA)CBCANH NMR experiment and (b) the 5D APSY-HC(CC-TOCSY)CONH experiments. The *dashed gray arrows* indicate INEPT magnetization transfer steps [64]. The *thick gray line* in (b) represents isotropic mixing. Adapted with permission from [63] and [65]

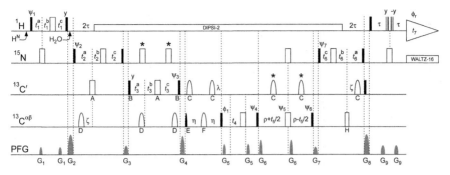

Fig. 8 Pulse sequence of the 7D APSY-HNCO(CA)CBCANH experiment. Radio-frequency pulses are applied at 118.0 ppm for ^{15}N, 173.0 ppm for ^{13}C', and at 42.0 ppm for ^{13}C$^{\alpha}$ and ^{13}C$^{\beta}$. At the start of each transient, the ^{1}H carrier frequency is set at 8.24 ppm, indicated by "H^N," and at the time point "H$_2$O" the carrier is changed to 4.7 ppm. *Black* and *white symbols* represent 90°- and 180°-pulses, respectively. *Unlabeled bars* stand for rectangular pulses applied at maximum power. Pulses *marked with capital letters* have individually adjusted lengths and shapes, depending on their purpose. All pulse lengths are given for a ^{1}H frequency of 750 MHz. ^{13}C'-pulses: A, 180°, rectangular shape, 38.3 μs; B, 90°, rectangular shape, 42.8 μs; C, 180°, I-Burp [66], 220 μs. ^{13}C$^{\alpha\beta}$-pulses: D, 180°, I-Burp (applied at 51.0 ppm), 220 μs; E, 90°, Gaussian cascade Q5 [67], 280 μs; F, 180°, Gaussian cascade Q3 [67], 185 μs; H, 180°, rectangular shape, 38.3 μs. The ^{15}N-pulses labeled with an *asterisk* are centered with respect to $t_3^a + t_3^b$ and t_3^c, respectively. The ^{13}C'-pulses labeled with an *asterisk* are centered with respect to $\rho + t_5/2$ and $\rho - t_5/2$, respectively. The last six pulses on the ^{1}H line represent a 3–9–19 Watergate pulse train [68]. Decoupling using DIPSI 2 [69] on ^{1}H and WALTZ-16 [70] on ^{15}N is indicated by *white rectangles*. The *triangle* with t_7 represents the acquisition period. On the *line marked PFG*, *curved shapes* indicate sine bell-shaped, pulsed magnetic field gradients along the z-axis with the following durations and strengths: G$_1$, 600 μs, 13 G/cm; G$_2$, 1,000 μs, 37 G/cm; G$_3$, 800 μs, 16 G/cm; G$_4$, 800 μs, 34 G/cm; G$_5$, 600 μs, 19 G/cm; G$_6$, 600 μs, 27 G/cm; G$_7$, 800 μs, 13 G/cm; G$_8$, 1,000 μs, 37 G/cm; G$_9$, 800 μs, 16 G/cm. Pulse phases different from x are indicated above the pulses. Phase cycling: $\phi_1 = \psi_2 = \phi_r = \{x, -x\}$, $\psi_4 = \{x, x, -x, -x\}$, $\psi_6 = y$. The initial delays were $t_1^a = t_1^c = 2.7$ ms, $t_2^a = t_2^c = 14.0$ ms, $t_3^a = t_3^c = 4.7$ ms, $t_6^a = t_6^c = 14.0$ ms, and $t_1^b = t_2^b = t_3^b = t_4 = t_5 = t_6^b = 0$ ms. Further delays were $\tau = 2.7$ ms, $\zeta = 14.0$ ms, $\eta = 6.8$ ms, $\lambda = 4.7$ ms, and $\rho = 20.75$ ms. Quadrature detection for the indirect dimensions was achieved using the trigonometric addition theorem [11, 15] with the phases ψ_1, ψ_2, ψ_3, ψ_4–ψ_6, ψ_6, and ψ_7 for t_1, t_2, t_3, t_4, t_5, and t_6, respectively. Evolution periods were implemented as direct evolution for t_4, and in constant-time fashion for t_1, t_2, t_3, t_5, and t_6. For t_1, t_2, t_3, and t_6, semi-constant time evolution was used for those maximal evolution periods that are too long to be accommodated in constant-time periods. Reproduced with permission from [63]

the residues 98–101, a backbone segment that features unfavorable backbone dynamics.

As for the 6D experiment, high precision of the chemical shift measurements is crucial for the sequential assignments, since these rely on matching of the amide ^{15}N and ^{1}H chemical shifts of sequentially neighboring amide moieties. A precision of 0.46 Hz and 0.44 Hz, respectively for the ^{1}H and ^{15}N chemical shifts was achieved (Fig. 9). Figure 9 also illustrates the significance of peak separation compared to the precision of the chemical shift measurements for unfolded proteins. This high precision enabled automated NMR assignment with the program

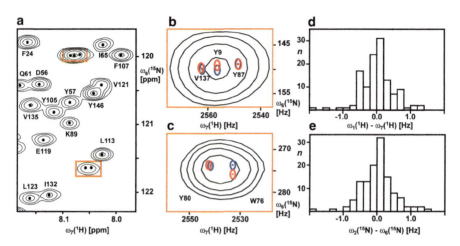

Fig. 9 Precision of chemical shift measurements by the 7D APSY-seq-HNCO(CA)CBCANH experiment. The data shown was recorded with a 3 mM sample of urea-denatured OmpX in 8 M urea aqueous solution at pH 6.5. (**a**) Spectral region from the (0°,0°,0°,0°,0°)-projection, which corresponds to a 2D [^{15}N,^{1}H]-correlation spectrum. The *black dots* are projections of the 7D peak positions determined by GAPRO, as represented by the [ω_6, ω_7]-correlations onto the experimental 2D projection spectrum. *Orange squares* indicate two clusters of overlapped signals which are displayed on an expanded scale in (**b**) and (**c**). (**b, c**) The two different [^{15}N,^{1}H]-pairs contained in each 7D signal are indicated in *red* ([ω_2, ω_1]-correlation) and *blue* ([ω_6, ω_7]-correlation). Contours are drawn at a distance of 1.0 Hz around the peak positions projected from the 7D data set. In (**a**)–(**c**), resonance assignments are given using one-letter amino acid symbols and the sequence positions. (**d**) Histogram of the variance between the measurements of the same amide proton chemical shift from the two 7D signals correlating two sequentially neighboring groups of 7 atoms. (**e**) Same as (**d**) for amide nitrogen-15 shifts. Reproduced with permission from [63]

GARANT [49] in spite of residual chemical shift degeneracy in some of the seven dimensions [63].

Since the longitudinal and transverse relaxation time constants in soluble non-globular proteins are in first order independent of the length of the polypeptide, similar experimental sensitivities can be expected for much larger unfolded proteins. Further, since the high precision of the APSY experiment of below 1 Hz falls substantially below the occurring distances between pairs of neighboring resonances in high-dimensional spaces [1, 72], similar assignment results to those achieved with the 150 residue OmpX can be expected for other unfolded proteins of much larger size. This has indeed been shown, where, by using the 7D APSY-seq-HNCO(CA)CBCANH experiment, the Zweckstetter group could assign the backbone resonances of the 441 residue Tau, a disordered polypeptide, within 5 days of measurement time, reducing the overall analysis time by more than order of magnitude as compared to a conventional approach [73]. APSY-NMR thus has tremendous potential for new insights into structure–function correlations of natively unfolded proteins, as well as for key contributions to the protein folding problem.

4.5 Automated NMR Assignment of Protein Side Chain Resonances

The precise APSY peak lists can also be the basis for side chain resonance assignments of proteins. A well suited magnetization transfer pathway for this purpose is the HC(CC-TOCSY)CONH pathway, which correlates side chain with backbone nuclei [74–78]. With APSY, the dimensionality of this experiment can be extended to five [65] (Fig. 7b). The pathway starts simultaneously on all aliphatic side chain protons including the H^α. An INEPT element transfers magnetization to the covalently bound carbon. Subsequently, the magnetization is transferred among the aliphatic carbon nuclei by isotropic mixing. At the end of the mixing time the magnetization on the C^α nucleus is transferred via the carbonyl carbon to the amide nitrogen of the successive amino acid residue and finally to the attached amide proton, from which the signal is acquired. For a given amino acid, the resulting 5D APSY correlation peak list thus contains a group of C–H correlations which have identical chemical shifts in the three backbone dimensions $\omega_3(^{13}C')$, $\omega_4(^{15}N)$, and $\omega_5(^1H^N)$.

The peak intensities of the correlation peaks in CC-TOCSY experiments depend strongly on the amplitude of the magnetization transfer during the isotropic mixing period and hence on the length of this period [79, 80]. There is no single mixing time for which all C–H moieties of all 20 amino acids have sufficiently large transfer amplitudes. This problem can be elegantly circumvented with APSY, since the TOCSY mixing time can be varied along with the projection angles. The analysis of the set of projection spectra with GAPRO does not require that a given 5D peak is present in all projections. By using a set of mixing times that enables sufficiently high transfer for all aliphatic side chain carbon moieties in some of the projections, it is possible to cover the resonance frequencies of all C–H moieties from all 20 amino acids in the resulting APSY correlation peak list.

Calculations of the transfer amplitudes in CC-TOCSY experiments show that the set of three mixing times – 12 ms, 18 ms, and 28 ms – covers all protons in the 20 amino acids [65]. The mixing time of 18 ms, which is commonly used in classical experiments, transfers magnetization from a majority of carbons in the side chains to the α-carbon nuclei. The mixing time of 12 ms is favorable for signals which have a small transfer at 18 ms. The long mixing time of 28 ms favors signals of long side chains, but also signals of short side chains, which are very weak or not present at the two other mixing times.

With these three mixing times, the 5D APSY-HC(CC-TOCSY)CONH experiment was recorded with a 1 mM solution of the 12.4-kDa globular protein TM1290 in 24 h of spectrometer time using 36 projections (Fig. 10). Based on the reference assignment of this protein, 424 cross peaks are expected in the resulting 5D APSY correlation peak list [43]; 368 thereof were actually found in the present experiment. These 368 correlations contained the chemical shifts of 97% of the aliphatic carbons and 87% of the aliphatic protons in the protein.

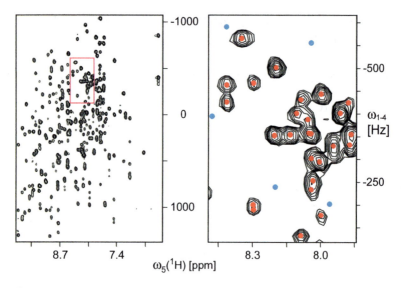

Fig. 10 2D projection of the 5D ASPY-HC(CC-TOCSY)CONH experiment with TM1290 recorded on a 750 MHz spectrometer using a TOCSY mixing time of = 17.75 ms. The projection with angles $\alpha = -46.6°$, $\beta = 0°$, $\gamma = -17.2°$ is shown. The region in *magenta* in the *left panel* is shown enlarged on the *right hand side*. The *colored dots* are the projections of the final 5D APSY peak list. *Red dots* indicate peaks present at the mixing time $\tau_m = 17.75$ ms, *blue dots* indicate peaks only present in spectra with other mixing times. Reproduced with permission from [65]

The resulting 5D APSY-HC(CC-TOCSY)CONH chemical shift correlation list together with the known backbone assignment are the sole input for the side chain assignment algorithm ALASCA (Algorithm for Local and linear Assignment of Side Chains from APSY data) [65]. In the ALASCA algorithm, each 5D APSY-HC(CC-TOCSY)CONH correlation is attributed to the residue, which has the nearest backbone chemical shifts in the 3D space of the ($\omega(^{13}C')$, $\omega(^{15}N)$, $\omega(^{1}H^N)$) frequencies. Subsequently, for each amino acid in the protein, the correlations of the TOCSY peak group are assigned to the side chain atoms by matching the chemical shifts of the 5D correlations to statistical values from the BMRB database [39].

As for the applications providing backbone assignments, the precision of the chemical shifts obtained for $\omega_5(^{1}H^N)$, $\omega_4(^{15}N)$, and $\omega_3(^{13}C')$ from the 5D APSY-HC (CC-TOCSY)CONH experiment is crucial for the assignment. It was found to be 0.5 Hz for $\omega_5(^{1}H^N)$, 2.3 Hz for $\omega_4(^{15}N)$, and 3.6 Hz for $\omega_3(^{13}C')$, which is substantially below the digital resolution of the individual projection spectra. With ALASCA all 368 peaks contained in the 5D peak list of TM1290 were correctly assigned.

The 5D APSY-HC(CC-TOCSY)CONH experiment was also used to assign the side chains of the two larger proteins, the 22-kDa protein kRas at 0.4 mM concentration and the 15-kDa drug target protein A at 0.3 mM concentration [51]. A total of 34 projections were recorded in 51 h on a 600 MHz spectrometer equipped with

Automated Projection Spectroscopy and Its Applications

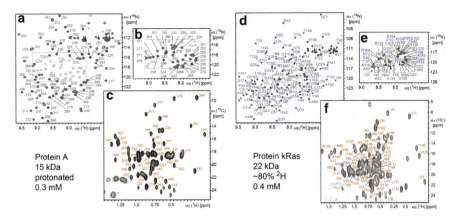

Fig. 11 Sequence-specific resonance assignments of two proteins obtained with the APSY CA-CB-CM strategy [51]. Data is shown for two proteins, the 15-kDa "protein A" (panels **a–c**) and the 22-kDa protein kRas (panels **d–f**). Selected sample parameters are indicated. The amide resonance assignments are shown in *blue* on 2D [^{15}N,^{1}H]-HSQC spectra (panels **a** and **d**), with their central regions shown enlarged (panels **b** and **e**). The methyl group assignments are indicated in *orange* on 2D [^{13}C,^{1}H]-HMQC spectra (panels **c** and **f**). Adapted with permission from [51]

a cryogenic probe at an experiment temperature of 23 °C. The assignment yielded for each protein nearly 90% of the Ala, Ile, Leu, Thr, and Val side chain methyl groups (Fig. 11).

Overall, the high quality of the GAPRO peak list of the 5D APSY-HC(CC-TOCSY)CONH experiment in terms of dimensionality, completeness, precision and very low number of artifacts provides an excellent basis for a reliable automated assignment of aliphatic side-chain atoms. Although the TOCSY mixing does not provide information on the direct covalent connectivities among the carbon nuclei, the 5D peaks can be used for reliable sequence-specific resonance assignment of aliphatic resonances, due to the availability of all five dimensions.

5 Conclusion and Outlook

This chapter presented the foundations of automated projection spectroscopy (APSY) that uses the algorithm GAPRO for automated spectral analysis. We showed applications of APSY for high-dimensional heteronuclear correlation NMR experiments with proteins. Without human intervention after the initial set-up of the experiments, complete peak lists for 4D to 7D NMR spectra, with a chemical shift precision of below 1 Hz, are typically obtained.

The positions of the peaks in the projection spectra that arise from a real N-dimensional peak are correlated among the projection spectra, whereas the positions of random noise are uncorrelated. This different behavior efficiently discriminates projected peaks against artifacts, and artifacts are therefore unlikely to appear in the final peak list. APSY is also well prepared to deal with inaccurate peak positions. Since the final N-dimensional APSY peak list is computed as the average of a large number of independent measurements, inaccurate peak positions in some of the projections have only a small influence on the overall precision. APSY has the advantage of relying exclusively on the analysis of experimental low-dimensional projection spectra, with no need ever to reconstruct the parent high-dimensional spectrum. APSY does not impose restrictions on the selection of the number of projections or the combinations of projection angles. The experience from our work indicates that sensitivity for signal detection rather than overcrowding of the 2D projection spectra is the limiting factor in practical applications of APSY-NMR with proteins.

In addition to providing automated peak picking and computation of the corresponding chemical shift lists, APSY supports automated sequential resonance assignment. Thus, APSY is a valid alternative to related NMR techniques. APSY can be the first step, after sample preparation, in a fully automated process of protein structure determination by NMR with successive automated algorithms for the NOESY spectrum analysis and structure calculation.

APSY software and tools can be downloaded from www.apsy.ch.

Acknowledgments Financial support from the Swiss National Science Foundation, the ETH Zürich, the NCCR Structural Biology, and the Biozentrum Basel is gratefully acknowledged.

References

1. Wüthrich K (1986) NMR of proteins and nucleic acids. Wiley, New York
2. Bax A, Grzesiek S (1993) Methodological advances in protein NMR. Acc Chem Res 26:131–138
3. Kay LE, Gardner KH (1997) Solution NMR spectroscopy beyond 25 kDa. Curr Opin Struct Biol 7:722–731
4. Wüthrich K (2003) NMR studies of structure and function of biological macromolecules (Nobel lecture). Angew Chem Int Ed 42:3340–3363
5. Ernst RR, Bodenhausen G, Wokaun A (1987) Principles of nuclear magnetic resonance in one and two dimensions. Oxford University Press, Oxford
6. Szyperski T, Yeh DC, Sukumaran DK, Moseley HNB, Montelione GT (2002) Reduced dimensionality NMR spectroscopy for high-throughput protein resonance assignment. Proc Natl Acad Sci USA 99:8009–8014
7. Orekhov VY, Ibraghimov I, Billeter M (2003) Optimizing resolution in multidimensional NMR by three-way decomposition. J Biomol NMR 27:165–173
8. Kozminski W, Zhukov I (2003) Multiple quadrature detection in reduced-dimensionality experiments. J Biomol NMR 26:157–166
9. Rovnyak D, Frueh DP, Sastry M, Sun ZYJ, Stern AS, Hoch JC, Wagner G (2004) Accelerated acquisition of high resolution triple-resonance spectra using non-uniform sampling and maximum entropy reconstruction. J Magn Reson 170:15–21
10. Szyperski T, Wider G, Bushweller JH, Wüthrich K (1993) Reduced dimensionality in triple-resonance NMR experiments. J Am Chem Soc 115:9307–9308
11. Brutscher B, Morelle N, Cordier F, Marion D (1995) Determination of an initial set of NOE-derived distance constraints for the structure determination of $^{15}N/^{13}C$-labeled proteins. J Magn Reson B 109:238–242

12. Kim S, Szyperski T (2003) GFT NMR, a new approach to rapidly obtain precise high-dimensional NMR spectral information. J Am Chem Soc 125:1385–1393
13. Kupce E, Freeman R (2003) Projection-reconstruction of three-dimensional NMR spectra. J Am Chem Soc 125:13958–13959
14. Kupce E, Freeman R (2003) Reconstruction of the three-dimensional NMR spectrum of a protein from a set of plane projections. J Biomol NMR 27:383–387
15. Kupce E, Freeman R (2004) Projection-reconstruction technique for speeding up multidimensional NMR spectroscopy. J Am Chem Soc 126:6429–6440
16. Kupce E, Freeman R (2004) Fast reconstruction of four-dimensional NMR spectra from plane projections. J Biomol NMR 28:391–395
17. Bracewell RN (1956) Strip integration in radio astronomy. Aust J Phys 9:198–217
18. Nagayama K, Bachmann P, Wüthrich K, Ernst RR (1978) The use of cross-sections and of projections in two-dimensional NMR spectroscopy. J Magn Reson 31:133–148
19. Mersereau RM, Oppenheim AV (1974) Digital reconstruction of multidimensional signals from their projections. Proc IEEE 62:1319–1338
20. Lauterbur PC (1973) Image formation by induced local interactions – examples employing nuclear magnetic resonance. Nature 242:190–191
21. Moseley HNB, Riaz N, Aramini JM, Szyperski T, Montelione GT (2004) A generalized approach to automated NMR peak list editing: application to reduced-dimensionality triple resonance spectra. J Magn Reson 170:263–277
22. Freeman R, Kupce E (2004) Distant echoes of the accordion: Reduced dimensionality, GFT-NMR, and projection-reconstruction of multidimensional spectra. Concepts Magn Reson A 23:63–75
23. Eghbalnia HR, Bahrami A, Tonelli M, Hallenga K, Markley JL (2005) High-resolution iterative frequency identification for NMR as a general strategy for multidimensional data collection. J Am Chem Soc 127:12528–12536
24. Luan T, Jaravine V, Yee A, Arrowsmith CH, Orekhov VY (2005) Optimization of resolution and sensitivity of 4D NOESY using multi-dimensional decomposition. J Biomol NMR 33:1–14
25. Malmodin D, Billeter M (2005) Multiway decomposition of NMR spectra with coupled evolution periods. J Am Chem Soc 127:13486–13487
26. Kupce E, Freeman R (2006) Hyperdimensional NMR spectroscopy. J Am Chem Soc 128:6020–6021
27. Szyperski T, Atreya HS (2006) Principles and applications of GFT projection NMR spectroscopy. Magn Reson Chem 44:S51–S60
28. Coggins BE, Zhou P (2006) PR-CALC: a program for the reconstruction of NMR spectra from projections. J Biomol NMR 34:179–195
29. Lescop E, Brutscher B (2007) Hyperdimensional protein NMR spectroscopy in peptide-sequence space. J Am Chem Soc 129:11916–11917
30. Cornilescu G, Bahrami A, Tonelli M, Markley JL, Eghbalnia HR (2007) HIFI-C: a robust and fast method for determining NMR couplings from adaptive 3D to 2D projections. J Biomol NMR 38:341–351
31. Mishkovsky M, Kupce E, Frydman L (2007) Ultrafast-based projection-reconstruction three-dimensional nuclear magnetic resonance spectroscopy. J Chem Phys 127:034507
32. Zhang Q, Atreya HS, Kamen DE, Girvin ME, Szyperski T (2008) GFT projection NMR based resonance assignment of membrane proteins: application to subunit C of $E.\ coli$ F_1F_0 ATP synthase in LPPG micelles. J Biomol NMR 40:157–163
33. Jaravine VA, Zhuravleva AV, Permi P, Ibraghimov I, Orekhov VY (2008) Hyperdimensional NMR spectroscopy with nonlinear sampling. J Am Chem Soc 130:3927–3936
34. Koradi R, Billeter M, Engeli M, Güntert P, Wüthrich K (1998) Automated peak picking and peak integration in macromolecular NMR spectra using AUTOPSY. J Magn Reson 135:288–297
35. Herrmann T, Güntert P, Wüthrich K (2002) Protein NMR structure determination with automated NOE-identification in the NOESY spectra using the new software ATNOS. J Biomol NMR 24:171–189

36. Baran MC, Huang YJ, Moseley HNB, Montelione GT (2004) Automated analysis of protein NMR assignments and structures. Chem Rev 104:3541–3555
37. Xia YL, Zhu G, Veeraraghavan S, Gao XL (2004) (3,2)D GFT-NMR experiments for fast data collection from proteins. J Biomol NMR 29:467–476
38. Hiller S, Fiorito F, Wüthrich K, Wider G (2005) Automated projection spectroscopy (APSY). Proc Natl Acad Sci USA 102:10876–10881
39. Seavey BR, Farr EA, Westler WM, Markley JL (1991) A relational database for sequence-specific protein NMR data. J Biomol NMR 1:217–236
40. Hiller S, Wider G, Wüthrich K (2008) APSY-NMR with proteins: practical aspects and backbone assignment. J Biomol NMR 42:179–195
41. Sattler M, Schleucher J, Griesinger C (1999) Heteronuclear multidimensional NMR experiments for the structure determination of proteins in solution employing pulsed field gradients. Prog Nucl Magn Reson Spectrosc 34:93–158
42. Wider G (1998) Technical aspects of NMR spectroscopy with biological macromolecules and studies of hydration in solution. Prog Nucl Magn Reson Spectrosc 32:193–275
43. Etezady-Esfarjani T, Peti W, Wüthrich K (2003) NMR assignment of the conserved hypothetical protein TM1290 of *Thermotoga maritima*. J Biomol NMR 25:167–168
44. Wider G, Dreier L (2006) Measuring protein concentrations by NMR spectroscopy. J Am Chem Soc 128:2571–2576
45. Richarz R, Wüthrich K (1978) ^{13}C NMR chemical shifts of common amino acid residues measured in aqueous solutions of linear tetrapeptides H–Gly–Gly–X–L-Ala–OH. Biopolymers 17:2133–2141
46. Oh BH, Westler WM, Darba P, Markley JL (1988) Protein ^{13}C spin systems by a single two-dimensional NMR experiment. Science 240:908–911
47. Grzesiek S, Bax A (1993) Amino acid type determination in the sequential assignment procedure of uniformly ^{13}C/^{15}N-enriched proteins. J Biomol NMR 3:185–204
48. Güntert P, Salzmann M, Braun D, Wüthrich K (2000) Sequence-specific NMR assignment of proteins by global fragment mapping with the program MAPPER. J Biomol NMR 18:129–137
49. Bartels C, Güntert P, Billeter M, Wüthrich K (1997) GARANT – a general algorithm for resonance assignment of multidimensional nuclear magnetic resonance spectra. J Comput Chem 18:139–149
50. Fiorito F, Hiller S, Wider G, Wüthrich K (2006) Automated resonance assignment of proteins: 6D APSY-NMR. J Biomol NMR 35:27–37
51. Gossert AD, Hiller S, Fernández C (2011) Automated NMR resonance assignment of large proteins for protein–ligand interaction studies. J Am Chem Soc 133:210–213
52. Anfinsen CB (1973) Principles that govern folding of protein chains. Science 181:223–230
53. Shortle D (1993) Denatured states of proteins and their roles in folding and stability. Curr Opin Struct Biol 3:66–74
54. Plaxco KW, Gross M (1997) Cell biology – the importance of being unfolded. Nature 386:657–659
55. Dunker AK, Lawson JD, Brown CJ, Williams RM, Romero P, Oh JS, Oldfield CJ, Campen AM, Ratliff CR, Hipps KW, Ausio J, Nissen MS, Reeves R, Kang CH, Kissinger CR, Bailey RW, Griswold MD, Chiu M, Garner EC, Obradovic Z (2001) Intrinsically disordered protein. J Mol Graph Model 19:26–59
56. Daggett V, Fersht AR (2003) Is there a unifying mechanism for protein folding? Trends Biochem Sci 28:18–25
57. Mayor U, Guydosh NR, Johnson CM, Grossmann JG, Sato S, Jas GS, Freund SMV, Alonso DOV, Daggett V, Fersht AR (2003) The complete folding pathway of a protein from nanoseconds to microseconds. Nature 421:863–867
58. Dyson HJ, Wright PE (2005) Intrinsically unstructured proteins and their functions. Nat Rev Mol Cell Biol 6:197–208
59. Lindorff-Larsen K, Rogen P, Paci E, Vendruscolo M, Dobson CM (2005) Protein folding and the organization of the protein topology universe. Trends Biochem Sci 30:13–19

60. Wüthrich K (1994) NMR assignments as a basis for structural characterization of denatured states of globular proteins. Curr Opin Struct Biol 4:93–99
61. Schwalbe H, Fiebig KM, Buck M, Jones JA, Grimshaw SB, Spencer A, Glaser SJ, Smith LJ, Dobson CM (1997) Structural and dynamical properties of a denatured protein. Heteronuclear 3D NMR experiments and theoretical simulations of lysozyme in 8 M urea. Biochemistry 36:8977–8991
62. Dyson HJ, Wright PE (2001) Nuclear magnetic resonance methods for elucidation of structure and dynamics in disordered states. Methods Enzymol 339:258–270
63. Hiller S, Wasmer C, Wider G, Wüthrich K (2007) Sequence-specific resonance assignment of soluble nonglobular proteins by 7D APSY-NMR spectroscopy. J Am Chem Soc 129:10823–10828
64. Morris GA, Freeman R (1979) Enhancement of NMR signals by polarization transfer. J Am Chem Soc 101:760–762
65. Hiller S, Joss R, Wider G (2008) Automated NMR assignment of protein side chain resonances using automated projection spectroscopy (APSY). J Am Chem Soc 130:12073–12079
66. Geen H, Freeman R (1991) Band-selective radiofrequency pulses. J Magn Reson 93:93–141
67. Emsley L, Bodenhausen G (1990) Gaussian pulse cascades – new analytical functions for rectangular selective inversion and in-phase excitation in NMR. Chem Phys Lett 165:469–476
68. Piotto M, Saudek V, Sklenar V (1992) Gradient-tailored excitation for single-quantum NMR spectroscopy of aqueous solutions. J Biomol NMR 2:661–665
69. Shaka AJ, Lee CJ, Pines A (1988) Iterative schemes for bilinear operators – application to spin decoupling. J Magn Reson 77:274–293
70. Shaka AJ, Keeler J, Frenkiel T, Freeman R (1983) An improved sequence for broad-band decoupling – WALTZ-16. J Magn Reson 52:335–338
71. Tafer H, Hiller S, Hilty C, Fernández C, Wüthrich K (2004) Nonrandom structure in the urea-unfolded *Escherichia coli* outer membrane protein X (OmpX). Biochemistry 43:860–869
72. Braun D, Wider G, Wüthrich K (1994) Sequence-corrected ^{15}N random coil chemical shifts. J Am Chem Soc 116:8466–8469
73. Narayanan RL, Dürr UHN, Bibow S, Biernat J, Mandelkow E, Zweckstetter M (2010) Automatic assignment of the intrinsically disordered protein Tau with 441-residues. J Am Chem Soc 132:11906–11907
74. Montelione GT, Lyons BA, Emerson SD, Tashiro M (1992) An efficient triple resonance experiment using carbon-13 isotropic mixing for determining sequence-specific resonance assignments of isotopically enriched proteins. J Am Chem Soc 114:10974–10975
75. Logan TM, Olejniczak ET, Xu RX, Fesik SW (1992) Side chain and backbone assignments in isotopically labeled proteins from two heteronuclear triple resonance experiments. FEBS Lett 314:413–418
76. Grzesiek S, Anglister J, Bax A (1993) Correlation of backbone amide and aliphatic side-chain resonances in ^{13}C/^{15}N-enriched proteins by isotropic mixing of ^{13}C magnetization. J Magn Reson B 101:114–119
77. Jiang L, Coggins BE, Zhou P (2005) Rapid assignment of protein side chain resonances using projection-reconstruction of (4,3)D HC(CCO)NH and intra-HC(C)NH experiments. J Magn Reson 175:170–176
78. Sun ZY, Hyberts SG, Rovnyak D, Park S, Stern AS, Hoch JC, Wagner G (2005) High-resolution aliphatic side-chain assignments in 3D HCcoNH experiments with joint H-C evolution and non-uniform sampling. J Biomol NMR 32:55–60
79. Braunschweiler L, Ernst RR (1983) Coherence transfer by isotropic mixing – application to proton correlation spectroscopy. J Magn Reson 53:521–528
80. Clore GM, Bax A, Driscoll PC, Wingfield PT, Gronenborn AM (1990) Assignment of the side-chain ^{1}H and ^{13}C resonances of interleukin-1 beta using double- and triple-resonance heteronuclear three-dimensional NMR spectroscopy. Biochemistry 29:8172–8184

Data Sampling in Multidimensional NMR: Fundamentals and Strategies

Mark W. Maciejewski, Mehdi Mobli, Adam D. Schuyler, Alan S. Stern, and Jeffrey C. Hoch

Abstract Beginning with the introduction of Fourier Transform NMR by Ernst and Anderson in 1966, time domain measurement of the impulse response (free induction decay) consisted of sampling the signal at a series of discrete intervals. For compatibility with the discrete Fourier transform, the intervals are kept uniform, and the Nyquist theorem dictates the largest value of the interval sufficient to avoid aliasing. With the proposal by Jeener of parametric sampling along an indirect time dimension, extension to multidimensional experiments employed the same sampling techniques used in one dimension, similarly subject to the Nyquist condition and suitable for processing via the discrete Fourier transform. The challenges of obtaining high-resolution spectral estimates from short data records were already well understood, and despite techniques such as linear prediction extrapolation, the achievable resolution in the indirect dimensions is limited by practical constraints on measuring time. The advent of methods of spectrum analysis capable of processing nonuniformly sampled data has led to an explosion in the development of novel sampling strategies that avoid the limits on resolution and measurement time imposed by uniform sampling. In this chapter we review the fundamentals of uniform and nonuniform sampling methods in one- and multidimensional NMR.

Keywords Nonuniform sampling · Spectrum analysis

M.W. Maciejewski, A.D. Schuyler, and J.C. Hoch (✉)
Department of Molecular, Microbial, and Structural Biology, University of Connecticut Health Center, 263 Farmington Ave, Farmington, CT 06030-3305, USA
e-mail: hoch@uchc.edu

M. Mobli
Division of Chemistry and Structural Biology, Institute for Molecular Bioscience, The University of Queensland, St. Lucia, Brisbane, QLD 4072, Australia

A.S. Stern
Rowland Institute at Harvard, 100 Edwin H. Land Blvd., Cambridge, MA 02142, USA

Contents

1. Introduction 50
2. Fundamentals: Sampling in One Dimension 51
   - 2.1 Oversampling 52
   - 2.2 How Long Should One Sample? 52
3. Sampling in Multiple Dimensions 53
   - 3.1 Quadrature Detection in Multiple Dimensions 53
   - 3.2 Sampling Limited Regime 54
4. Non-Fourier Methods of Spectrum Analysis 54
   - 4.1 "DFT" of NUS Data and the Point-Spread Function 55
5. Nonuniform Sampling: A Brief History 59
   - 5.1 The Accordion 59
   - 5.2 Random Sampling 59
   - 5.3 RD, Redux 60
   - 5.4 The NUS Explosion 61
6. General Aspects of Nonuniform Sampling 62
   - 6.1 On-Grid vs Off-Grid Sampling 62
   - 6.2 Bandwidth and Aliasing 62
   - 6.3 Sampling Artifacts Are Spectral Aliases 62
7. A Menagerie of Sampling Schemes 65
   - 7.1 Random and Biased Random Sampling 66
   - 7.2 Triangular 69
   - 7.3 Radial 69
   - 7.4 Concentric Rings 70
   - 7.5 Spiral 70
   - 7.6 Beat-Matched Sampling 71
   - 7.7 Poisson Gap Sampling 72
   - 7.8 Burst Sampling 72
   - 7.9 Nonuniform Averaging 72
   - 7.10 Random Phase Detection 73
   - 7.11 Optimal Sampling? 73
8. Concluding Remarks 74
References 75

1 Introduction

Since the introduction of Fourier Transform NMR by Richard Ernst and Weston Anderson in 1966, the measurement of NMR spectra has principally involved the measurement of the free induction decay (FID) following the application of a broad-band RF pulse to the sample [1]. The FID is measured at regular intervals, and the spectrum obtained by computing the discrete Fourier transform (DFT). The accuracy of the spectrum obtained by this approach depends critically on how the data are sampled. In the application of this approach to multidimensional NMR experiments, the constraint of uniform sampling interval imposed by the DFT incurs substantial sampling burdens. The advent of non-Fourier methods of spectrum analysis that do not require data sampled at uniform intervals has enabled the development of a host of nonuniform sampling (NUS) strategies. In this chapter we review the fundamentals of sampling, both uniform and nonuniform, in one and multiple dimensions. We then survey recently developed NUS methods that have been applied to multidimensional NMR, and consider prospects for new developments. While non-Fourier methods of spectrum analysis are indispensible for nonuniformly sampled data, they have been reviewed elsewhere.

2 Fundamentals: Sampling in One Dimension

Implicit in the definition of the complex DFT,

$$f_n = \frac{1}{\sqrt{N}} \sum_{k=0}^{N-1} d_k e^{-2\pi i k n/N} \qquad (1)$$

is the periodicity of the spectrum, which is apparent by setting k to N in (1). Thus the component at frequency $n\Delta t/N$ is equivalent to (and indistinguishable from) the components at $(n/N\Delta t) +/- (m/\Delta t)$, $m = 1, 2, \ldots$. This periodicity makes it possible to consider the DFT spectrum as containing all positive frequencies with zero frequency at one edge, or containing both positive and negative frequencies with zero frequency at (actually near) the middle. The equivalence of frequencies in the DFT spectrum that differ by a multiple of $1/\Delta t$ is a manifestation of the Nyquist sampling theorem, which states that, in order to determine unambiguously the frequency of an oscillating signal from a set of uniformly spaced samples, the sampling interval must be at least $1/\Delta t$. (For additional details of the DFT and its application in NMR, see [2].)

In the description of the DFT given by (1) it is assumed that the data samples and DFT spectrum are both complex. Implicit in this description is that two orthogonal components of the signal are sampled at the same time, referred to as simultaneous quadrature detection. Most modern NMR spectrometers are capable of simultaneous quadrature detection, but early instruments had a single detector, so only a single component of the signal could be sampled at any one time. With so-called single-phase detection, the sign of the frequency is indeterminate. Consequently the carrier frequency must be placed at one edge of the spectral region and the data must be sampled at $1/2\Delta t$ to determine unambiguously the frequencies of signals spanning a bandwidth (or spectral width, SW) $1/\Delta t$.

The detection of two orthogonal components permits the sign ambiguity to be resolved while sampling at a rate of $1/\Delta t$. This approach, called phase sensitive or quadrature detection, enables the carrier to be placed at the center of the spectrum. Simultaneous quadrature detection is commonly achieved by mixing a detected signal with a fixed-frequency reference signal and the same reference signal phase shifted by 90°, or a cosinusoidal reference. The output of the phase-sensitive detector is two signals, differing in phase by 90°, containing frequency components of the original signal oscillating at the sum and difference of the reference frequency with the original frequencies. The sum frequencies are typically filtered out using a low-pass filter. While quadrature detection enables the sign of frequencies to be determined unambiguously, while sampling at $1/\Delta t$, it requires just as many data samples as single-phase detection since it samples the signal twice at each $1/\Delta t$ interval, while single-phase detection samples once at each $1/2\Delta t$ interval.

2.1 Oversampling

The Nyquist theorem places a lower bound on the sampling rate, but what about sampling faster? It turns out that sampling faster than the reciprocal of the spectral width, called oversampling, can provide some benefits. One is that the oversampling increases the dynamic range, the ratio between the largest and smallest (non-zero) signals that can be detected [3, 4]. Analog-to-digital (A/D) converters employed in most NMR spectrometers represent the converted signal with fixed binary precision, e.g., 14 or 16 bits. A 16-bit A/D converter can represent signed integers between $-32,768$ and $+32,767$. Oversampling by a factor of n effectively increases the dynamic range by sqrt(n). Another benefit of oversampling is that it prevents certain sources of noise that are NOT band-limited to the same extent as the systematic (NMR) signals from being aliased into the spectral window.

2.2 How Long Should One Sample?

For signals that are stationary, that is their behavior doesn't change with time, the longer you sample the better the sensitivity and accuracy. For normally distributed random noise, the signal-to-noise (S/N) ratio improves with the square root of the number of samples. NMR signals are rarely stationary, however, and the signal envelope typically decays exponentially in time. For decaying signals, there invariably comes a time when collecting additional samples is counter-productive, because the amplitude of the signal has diminished below the amplitude of the noise, and additional sampling only serves to reduce S/N. The time $1.3 \times R_2$, where R_2 is the decay rate of the signal, is the point of diminishing returns, beyond which additional data collection results in reduced sensitivity [5]. It makes sense to sample at least this long in order to optimize the sensitivity per unit time of an experiment. However, limiting sampling to $1.3 \times R_2$ results in a compromise. That's because the ability to distinguish signals that have similar frequencies increases the longer one samples. Intuitively this makes sense because the longer two signals with different frequencies evolve, the greater the difference in their values at a specific time. Thus resolution, the ability to distinguish closely-spaced frequency components, is largely related to the longest time sample.

3 Sampling in Multiple Dimensions

While the FTNMR experiment of Ernst and Anderson was the seminal development behind all of modern NMR spectroscopy, it wasn't until 1971 that Jean Jeener proposed a strategy for parametric sampling of a virtual or indirect time dimension that formed the basis for modern multidimensional NMR [6], including applications to magnetic resonance imaging (MRI). In the simplest realization, an indirect time dimension can be defined as the time between two RF pulses applied in an NMR experiment. The FID is recorded subsequent to the second pulse, and because it evolves in real time its evolution is said to occur in the acquisition dimension. A given experiment can only be conducted using a single value of the time interval between pulses, but the indirect time dimension can be explored by repeating the experiment using different values of the time delay. When the values of the time delay are systematically varied using a fixed sampling interval, the resulting spectrum as a function of the time interval can be computed using the DFT along the columns of the two-dimensional data matrix, with rows corresponding to samples in the acquisition dimension and columns the indirect dimension. Generalization of the Jeener principle to an arbitrary number of dimensions is straightforward, limited only by the imagination of the spectroscopist and the ability of the spin system to maintain coherence over an increasingly lengthy sequence of RF pulses and indirect evolution times.

3.1 Quadrature Detection in Multiple Dimensions

Left ambiguous in the discussion above of multidimensional NMR experiments is the problem of frequency sign discrimination in the indirect dimensions. Because the indirect dimensions are sampled parametrically, i.e., each indirect evolution time is sampled via a separate experiment, the possibility of simultaneous quadrature detection is not available. Quadrature detection in the indirect dimension of a two-dimensional experiment nonetheless can be accomplished by using two experiments for each indirect evolution time to determine two orthogonal responses. This approach was first described by States, Haberkorn, and Ruben, and is frequently referred to as the States method [7]. Alternatively, oversampling could be used by sampling at twice the Nyquist frequency while rotating the detector phase through 0°, 90°, 180°, and 270°, an approach called time-proportional phase incrementation (TPPI). A hybrid approach is referred to as States-TPPI. Processing of States-TPPI sampling is performed using a complex DFT, just as for States sampling, while TPPI employs a real-only DFT.

3.2 Sampling Limited Regime

An implication of the Jeener strategy for multidimensional experiments is that the length of time required to conduct a multidimensional experiment is directly proportional to the total number of indirect time samples (times two for each indirect dimension if States or States-TPPI sampling is used). In experiments that permit the spin system to return close to equilibrium by waiting on the order of T_1 before performing another experiment, sampling along the acquisition dimension effectively incurs no time cost. Sampling to the Rovnyak limit ($1.3 \times R_2$) in the indirect dimensions places a substantial burden on data collection, even for experiments on proteins with relatively short relaxation times. Thus a three dimensional experiment for a 20-kDa protein at 14 T (600 MHz for ^{1}H) exploring ^{13}C and ^{15}N frequencies in the indirect dimensions would require 2.6 days in order to sample to $1.3 \times R_2$ in both indirect dimensions. A comparable four-dimensional experiment with two ^{13}C (aliphatic and carbonyl) and one ^{15}N indirect dimensions would require 137 days. As a practical matter, multidimensional NMR experiments rarely exceed a week, as superconducting magnets typically require cryogen refill on a weekly basis. Thus multidimensional experiments rarely achieve the full potential of the resolution afforded by superconducting magnets. The problem becomes more acute with very high magnetic fields. The time required for data collection in a multidimensional experiment to fixed maximum evolution times in the indirect dimensions increases with the increase in magnetic field raised to the power of the number of indirect dimensions. The same four-dimensional protein NMR experiment performed at 21.2 T (900 MHz for ^{1}H), sampled to $1.3 \times R_2$, would require about 320 days. NUS approaches have made it possible to conduct high resolution 4D experiments that would otherwise be impractical [45].

While methods of spectrum analysis capable of super-resolution exist, that is, methods that can achieve resolution greater than $1/t_{max}$, the most common of these, linear prediction (LP) extrapolation, has substantial drawbacks. LP extrapolation is used to extrapolate signals beyond the measured interval. While this can dramatically suppress truncation artifacts associated with zero-filling as well as improve resolution, because LP extrapolation implicitly assumes exponential decay it can lead to subtle frequency bias when the signal decay is not perfectly exponential [8]. This bias can have detrimental consequences for applications that require the determination of small frequency differences, such as measurement of residual dipolar couplings (RDCs).

4 Non-Fourier Methods of Spectrum Analysis

The DFT, strictly speaking, requires data sampled at uniform intervals. Thus the development of NUS methods to avoid the sampling limited regime in multidimensional NMR closely parallels the development of non-Fourier methods of spectrum analysis capable of treating data that have been collected at nonuniform intervals. One of the first methods to be employed in NMR in conjunction with NUS is maximum entropy (MaxEnt) reconstruction [9, 10]. MaxEnt reconstruction seeks that spectrum containing the least amount of information that is still consistent with the measured data. It makes no assumption regarding the nature of the signal, and thus is suitable for application to signals characterized by non-exponential decay (non-Lorentzian line shapes). A host of similar methods employ functionals other than the entropy to regularize the spectrum, for example the l_1-norm [11, 12]. Another class of methods that can reconstruct frequency spectra from data that are sampled nonuniformly assume a model for the data. Bayesian [13] and maximum likelihood [14, 15] (MLM) methods both assume the signal can be described as a sum of exponentially decaying sinusoids, and can be used either to reconstruct a frequency spectrum or to determine a list of frequency components and their characteristics; for this reason these methods are often described as being *parametric*. A method that is intermediate between the parametric methods that assume a model for the signal and regularization methods that do not is a method called multidimensional decomposition [16] (MDD). It assumes that frequency components in multidimensional spectra can be decomposed into a vector product of one-dimensional lineshapes. The approach is related to principle component analysis, and has been utilized in the field of analytical chemistry and chemometrics (where it is called PARAFAC [17]); a unique decomposition exists only for spectra that have three or more dimensions.

4.1 *"DFT" of NUS Data and the Point-Spread Function*

From the definition of the DFT, it is clear that the Fourier sum can be modified by evaluating the summand at arbitrary frequencies rather than uniformly spaced frequencies. Kozminksi and colleagues have proposed utilizing this approach for computing frequency spectra of NUS data [18]; however, strictly speaking it is no longer properly called a Fourier transformation of the NUS data. Consider the special case where the summand in (1) is evaluated for a subset of the normal regularly-spaced time intervals. An important characteristic of the DFT is the orthogonality of the basic functions (the complex exponentials),

$$\sum_{n=0}^{N-1} e^{-2\pi i (k-k')n/N} = 0, \quad k \neq k', \qquad (2)$$

when the summation is carried out over a subset of the time intervals. Some of the values of n indicated by the sum in (2) are left out, and the complex exponentials are no longer orthogonal. An implication is that frequency components in the signal *interfere* with one another when the sampling is nonuniform.

Consider now NUS data sampled at the same subset of uniformly spaced times, but supplemented by the value zero for those times not sampled. Clearly the DFT can be applied to this augmented data, but it is not the same as "applying the DFT to NUS data." It is a subtle distinction but an important one. What is frequently referred to as the DFT spectrum of NUS data is not the spectrum of the NUS data but the spectrum of the *zero-augmented data*. The differences between the DFT of the zero-augmented data and the spectrum of the signal are mainly the result of the choice of sampled times, and are called sampling artifacts. While the DFT of zero-augmented data is not the spectrum we seek, it can sometimes be a useful approximation if the sampled times are chosen carefully to diminish the sampling artifacts.

The application of the DFT to NUS data has parallels in the problem of numerical quadrature on an irregular mesh, or evaluating an integral on a set of irregularly-spaced points [19]. The accuracy of the integral estimated from discrete samples is typically improved by judicious choice of the sample points, or pivots, and by weighting the value of the function being integrated at each of the pivots. For pivots (sampling schedules) that can be described analytically, the weights correspond to the Jacobian for the transformation between coordinate systems (as for the polar FT, discussed below). For sampling schemes that cannot be described analytically, for example those given with a random distribution, the Voronoi area (in two dimensions; volume in three dimensions, etc.) provides a useful set of weights [20]. The Voronoi area is the area occupied by the set of points around each pivot that are closer to that pivot than to any other pivot in the NUS set.

Under certain conditions the relationship between the DFT of the zero-augmented NUS data and the true spectrum has a particularly simple form. If the sampling is restricted to the uniformly-spaced Nyquist grid (also referred to as the Cartesian sampling grid) and there exists a real-valued sampling function with the property that when it multiplies a uniformly sampled data vector, element-wise, the result is the zero-augmented NUS data vector, then the DFT of the zero-augmented NUS data is the convolution of the DFT spectrum of the uniformly sampled data with the DFT of the sampling function. The sampling function has the value 1 for times that are sampled and zero for times that are not. The DFT of the sampling function is variously called the point-spread function (PSF), the impulse response, or the sampling spectrum.

The PSF provides insight into the locations and magnitudes of sampling artifacts that result from NUS, and it can have an arbitrary number of dimensions, corresponding to the number of dimensions in which NUS is applied. The PSF typically consists of a main central component at zero frequency, with smaller non-zero frequency components. Because the PSF enters into the DFT of the zero-augmented spectrum through convolution, each non-zero frequency component of the PSF will give rise to a sampling artifact for each component in the signal spectrum, with positions relative to the signal components that are the same as the relationship of the satellite peaks in the PSF. The amplitudes of the sampling artifacts will be proportional to the amplitude of the signal component and the relative height of the satellite peaks in the PSF. Thus the largest sampling artifacts will arise from the largest-amplitude components of the signal spectrum. The effective dynamic range (ratio between the magnitude of the largest and smallest signal component that can be unambiguously identified) of the DFT spectrum of the zero-augmented data can be directly estimated from the PSF for a sampling scheme as the ratio between the amplitude of the largest non-zero frequency component and the amplitude of the zero-frequency component.

Using NUS approaches to reconstruct a fully-dimensional spectrum invariably introduces sampling artifacts that are characteristic of the NUS strategy employed. Characteristic ridge artifacts emanating from peaks in back-projection reconstruction (BPR) spectra (described below) that were initially believed to be artifacts of back projection were instead demonstrated to be characteristic of radial sampling by using MaxEnt reconstruction to process radially-sampled data: the MaxEnt spectrum contained essentially identical ridge artifacts [44]. While spectral reconstruction methods attempt to diminish sampling artifacts in the reconstructed spectrum, their ability to suppress sampling artifacts is limited by the presence of noise. It is thus important to have an understanding of the nature of sampling artifacts that is independent of the method used to reconstruct the spectrum. Provided that sampling is restricted to a uniform Cartesian grid (arbitrary sampling schemes can be treated using successively fine grids) and one can define a real-valued sampling function that has the value one when a sample is collected and zero when it is not collected, sampling artifacts arise from the convolution of the impulse response or PSF with the true spectrum. The PSF is simply given by the DFT of the sampling function. PSFs typically exhibit a major peak at zero frequency, with satellite peaks of varying intensity at non-zero frequencies. Using the DFT to process NUS data, the resulting spectrum corresponds exactly to the convolution of the PSF with the true spectrum (Fig. 1). Methods such as MaxEnt reconstruction suppress the magnitude of sampling artifacts, but they appear at the same locations as found in the DFT spectrum (Fig. 2).

In addition to helping to specify the frequencies of sampling artifacts (which will depend on the frequencies contained in the signal being sampled as well as the sampling scheme), the PSF helps to specify the magnitudes of the sampling artifacts, as discussed above. While MaxEnt or other methods of spectrum analysis that attempt to deconvolve the PSF can improve the dynamic range, sampling schemes with PSFs containing smaller satellite peaks (relative to the central component) will give rise to smaller sampling artifacts.

An implication of restricting the sampling function to being a real vector is that if quadrature detection is employed in the indirect dimensions, e.g., States-Haberkorn-Ruben, then all quadrature components must be sampled for a given set of indirect evolution times. If they are not all sampled, the sampling function is complex, and the relationship between the DFT of the NUS data, the DFT of the sampling function, and the true spectrum is no longer a simple convolution.

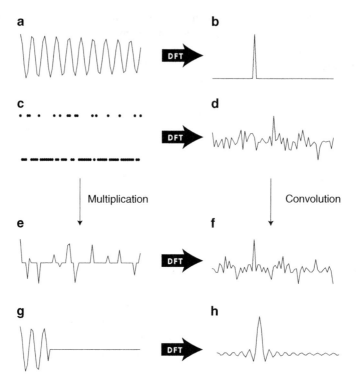

Fig. 1 The DFT of a decaying sinusoid (**a**, **b**) and a sampling function (**c**, **d**) and their multiplication in the time domain (**e**) resulting in their convolution in the frequency domain (**f**). The DFT of the sampling function (**f**) is the PSF

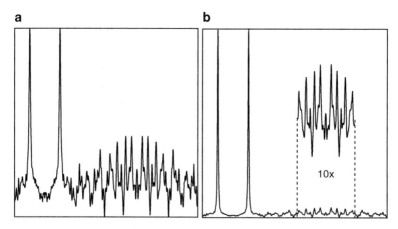

Fig. 2 (**a**) nuDFT vs (**b**) MaxEnt reconstruction applied to the same data. The *inset* in B shows a tenfold expansion of the baseline

5 Nonuniform Sampling: A Brief History

5.1 The Accordion

It was recognized soon after the development of FT NMR that one way to avoid the sampling limited regime in multidimensional situations is to avoid collecting the entire Nyquist grid in the indirect time dimensions. The principal challenge to this idea was that methods for computing the multidimensional spectrum from nonuniformly sampled data were not widely available. In 1981 Bodenhausen and Ernst introduced a means of avoiding the sampling constraints associated with uniform parametric sampling of two indirect dimensions of three-dimensional experiments, while also avoiding the need to compute a multidimensional spectrum from an incomplete data matrix, by coupling the two indirect evolution times [21]. By incrementing the evolution times in concert, sampling occurs along a radial vector in t_1-t_2, with a slope given by the ratio of the increments applied along each dimension. This effectively creates an aggregate evolution time $t = t_1 + \alpha*t_2$ that is sampled uniformly, and thus the DFT can be applied to determine the frequency spectrum. According to the projection-cross-section theorem, this spectrum is the projection of the full t_1-t_2 spectrum onto a vector with angle α in the f_1-f_2 plane. Bodenhausen and Ernst referred to this as an "accordion" experiment. Although they did not propose reconstruction of the full f_1-f_2 spectrum from multiple projections, they did discuss the use of multiple projections for characterizing the corresponding f_1-f_2 spectrum, and thus the accordion experiment is the precursor to more recent radial sampling methods that are discussed below. Because the coupling of evolution times effectively combines time (and the corresponding frequency) dimensions, the accordion experiment is an example of a reduced dimensionality (RD) experiment.

5.2 Random Sampling

The 3D accordion experiment has much lower sampling requirements because it avoids sampling the Cartesian grid of (t_1, t_2) values that must be sampled in order to utilize the DFT to compute the spectrum along both t_1 and t_2. A more general approach than the accordion experiment is to eschew regular sampling altogether. A consequence of this approach is that one cannot utilize the DFT to compute the spectrum, so some alternative method capable of utilizing nonuniformly sampled data must be employed. In seminal work, Laue and collaborators introduced the use of MaxEnt reconstruction to compute the frequency spectrum from nonuniformly sampled data [22]. In analogy with the concept of matched filter apodization for maximizing signal-to-noise ratio (S/N), Barna et al. utilized random sampling that was exponentially biased to short times, so that the sampling distribution matched the decay of the signal envelope. The concept of biased random sampling was further generalized to J-modulated experiments (cosine-modulated exponential decay) and constant-time experiments (no decay) by Schmieder et al. [32, 33] While the combination of biased random sampling and MaxEnt reconstruction provided high resolution spectra with dramatic reductions in experiment time compared to conventional uniform sampling because it employs samples collected at long evolution times without the need to sample all uniformly-spaced shorter times, the approach was not widely adopted, no doubt because neither MaxEnt reconstruction nor NUS was highly intuitive. Nevertheless a small cadre of investigators continued to explore novel NUS schemes in conjunction with MaxEnt reconstruction throughout the 1990s.

5.3 RD, Redux

The first RD experiment was the accordion experiment. In the original accordion experiment one indirect dimension represented chemical shift evolution while the second indirect dimension encoded a mixing time designed to measure chemical exchange. Although this experiment established the foundation for a host of subsequent RD experiments, most of which deal exclusively with chemical shift evolution, its utility for measuring relaxation rates and other applications is still being developed [23, 24]. Even though it was clear from the initial description of the accordion experiment that the method was applicable to any 3D experiment, it was nearly a decade before it was applied to a 3D experiment where both indirect dimensions represented chemical shifts [25, 26]. This application emerged as a consequence of newly-developed methods for isotopic labeling of proteins that enabled multinuclear, multidimensional experiments, with reasonable sensitivity, for sequential resonance assignment and structure determination of proteins. The acquisition of two coupled frequency dimensions, however, introduces some difficulties. The main problem is that the two dimensions being co-evolved are mixed and must be deconvoluted before any useful information can be extracted. Since the evolution linearly combines the two dimensions, their frequencies are "mixed" in the spectrum in a linear manner as well. The number of resonances observed in the lower dimensional spectrum depends on the number of linked dimensions. Thus, if two dimensions are linked, the RD spectrum will contain two peaks per resonance of the higher dimensional spectrum, whereas if three dimensions are coupled, each of the above two peaks will be split by the second frequency resulting in four resonances and so on. The position of the peaks in the spectrum can be used to extract the true frequency of the resonances in the spectrum. The problem obviously becomes more complicated as the number of resonances is increased. If overlap can be avoided, however, it is possible to reduce experimental time drastically. Among the weaknesses of this approach are sensitivity losses, associated with both peak splitting and relaxation losses, which effectively limit the number of dimensions that can be coupled for a given molecular size.

An extension of RD was presented by Kim and Szyperski [27] in 2003 in which they used a "G-matrix" to combine appropriately the hypercomplex data of arbitrary dimensionality to produce "basic spectra." These spectra are much less complicated than the RD projections and the known relationship between the various patterns can be used to extract true chemical shifts (via nonlinear least-squares fitting). Combination of the hypercomplex planes enables recovery of some sensitivity that is otherwise lost in RD approaches due to peak splitting. A disadvantage is that the data are not combined in a higher dimensional spectrum so that the sensitivity is related to that of each of the lower dimensional projections rather than the entire dataset. GFT-NMR was developed contemporaneously with advances in sensitivity delivered by higher magnetic fields and cryogenically cooled probes, providing sufficient sensitivity to make GFT experiments feasible for the first time, albeit using very concentrated protein samples (the GFT method was demonstrated on a 2 mM sample of ubiquitin).

Broader appreciation for NUS was stimulated by a series of papers by Kupce and Freeman, in which they utilized BPR from a series of experiments employing radial sampling in t_1/t_2 to reconstruct the fully-dimensional $f_1/f_2/f_3$ spectrum [28–32]. While the data sampling was equivalent to that employed by the accordion, GFT, and RD experiments, the use of back-projection (by analogy to computerized tomography) demonstrated the connection with the 3D spectrum conventionally obtained by uniform sampling and DFT. Despite some drawbacks to radial sampling (discussed below), the BPR approach was important because it provided a useful heuristic for more general NUS approaches.

The principle underlying radial sampling in 3D experiments generalizes to higher dimensions. For example, coupling of three indirect evolution times results in a projection of three dimensions onto a vector with one angle specifying the orientation with respect to the t_1/t_2 plane, and one specifying the angle with respect to the t_2/t_3 plane. Two very similar approaches for circumventing sampling limitations associated with uniform sampling in higher-dimensional experiments have been introduced to achieve high resolution while employing prior knowledge to design sampling angles. Chemical shift distributions expected for proteins can be used to determine a set of radial sampling angles (projection angles) that will optimally resolve potential overlap. Identification of frequencies in the projected spectra, together with knowledge of the projection angles, can be used to determine the (unprojected) frequencies in the orthogonal coordinate system of the fully-dimensional experiment.

In addition to GFT and BPR, a host of other methods can be applied to radially-sampled data; like BPR, these methods reconstruct the fully-dimensional spectrum. Zhou and colleagues employed radial FT [28] to process data collected along concentric rings in t_1/t_2 [29]. MLM methods that fit a model (typically consisting of a sum of exponentially-damped sinusoids) can also be used to analyze radially sampled data, as can regularization methods that do not model the signal (e.g., l_1-norm, MaxEnt).

5.4 The NUS Explosion

Since the turn of the twenty-first century, there has been a great deal of effort devoted to developing novel NUS strategies for multidimensional NMR. A recurring theme has been the importance of irregularity or randomness. Approaches involving various analytic sampling schemes (triangular, concentric rings, spirals) as well as novel random distributions (Poisson gap) have been described.

6 General Aspects of Nonuniform Sampling

We will contrast different approaches to NUS that have been applied to multi-dimensional NMR in a moment, but we first discuss some characteristics of NUS that are general and apply to all NUS approaches.

6.1 On-Grid vs Off-Grid Sampling

NUS schemes are sometimes characterized as on-grid or off-grid. Schemes that sample a subset of the evolution times normally sampled using uniform sampling at the Nyquist rate (or faster) are called on-grid. In schemes such as radial, spiral, or concentric ring, the samples do not fall on the same Cartesian grid. However, one can define a Cartesian grid with spacing determined by the precision with which evolution times are specified (discussed below). Alternatively, "off grid" sampling schemes can be approximated by "aliasing" (this time in the computer graphics sense) the evolution times onto a Nyquist grid, without greatly impacting the sampling artifacts (Fig. 3).

6.2 Bandwidth and Aliasing

Bretthorst was the first to consider carefully the implications of NUS for bandwidth and aliasing [30, 31]; his important contribution was published in a rather obscure proceedings volume, but more recently a version has been published in a more widely-accessible publication. Among the major points Bretthorst raises is that sampling artifacts accompanying NUS can be viewed as aliases. This is demonstrated in Fig. 4, where the spectrum obtained using uniform but deliberate undersampling is contrasted with the DFT spectrum for NUS data of the same signal.

6.3 Sampling Artifacts Are Spectral Aliases

However, as Bretthorst has pointed out, when the sampling (evolution) times are specified with finite precision (as they are in all commercial instruments), one can always define a uniform grid on which all the samples will fall. This grid spacing will generally be finer than the Nyquist grid. The largest grid spacing sufficient to encompass fully an arbitrary NUS scheme is given by the greatest common

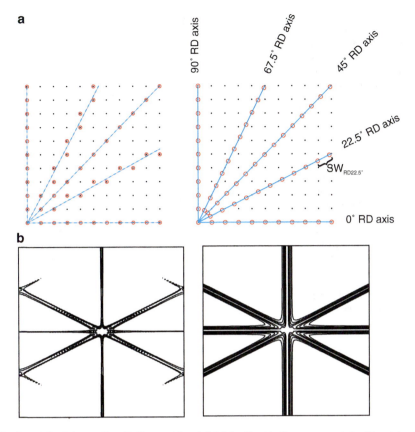

Fig. 3 (**a**) Radial sampling (*left*) on-grid and (*right*) off-grid. *Dots* represent the Nyquist grid, *circles* represent sampled data points. The *solid lines* indicate the angle of the radial vector (projection axis). (**b**) Reconstruction of radially sampled data; on-grid sampling reconstructed using MaxEnt (*left*) and off-grid sampled data reconstructed using PR (*right*)

divisor (GCD) of the sampled times, which is at least as large as the precision and may be larger, depending on the sampling scheme. As the samples are not uniform, the Nyquist sampling theorem does not apply, so strictly speaking there is not a bandwidth limiting the frequencies that can be unambiguously determined.

NUS artifacts are a form of aliasing, which can be appreciated by considering uniform undersampling as a form of NUS. Figure 4 illustrates a one-dimensional spectrum computed by applying the DFT to a synthetic signal sampled at the Nyquist interval (Fig. 4a) and twice the Nyquist interval (Fig. 4b). The signal sampled at twice the Nyquist interval has one alias of the true signal. Figure 4c depicts the DFT spectrum of a signal sampled nonuniformly. Note the strongest sampling artifact occurs precisely at the location of the undersampling artifact. Higher order sampling artifacts can be ascribed to aliases due to undersampling by greater degrees.

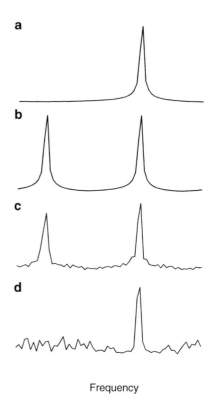

Fig. 4 Examples of aliasing using uniform (**a, b**) and nonuniform (**c, d**) sampling. (**a, b**) DFT spectrum using uniform sampling for a single synthetic sinusoid; (**a**) at the Nyquist rate, (**b**) at one-half the Nyquist rate. (**c, d**) nuDFT (DFT in which samples not measured are set to zero) for the same synthetic signal using nonuniform sampling from the Nyquist grid. In (**c**) an alias appears at the frequency expected using deliberate undersampling by a factor of 2, but with a height slightly less than the true (*unaliased*) peak. In (**d**) the alias is greatly diminished, a result of the greater number of samples in the NUS set spaced at the Nyquist interval

Since sampling artifacts are aliases, they can be diminished by increasing the effective bandwidth. One way to do this is to decrease the GCD. As shown above the GCD need not correspond to the spacing of the underlying grid. Introducing irregularity is one way to decrease the GCD to the size of the grid, and this helps to explain the usefulness of randomness for reducing artifacts from NUS schemes [46]. The ability of randomness to reduce NUS aliasing artifacts is depicted in Fig. 5. The left panels depict a two-dimensional sampling scheme (top) in which the data are undersampled by a factor of four in each dimension, leading to multiple instances of each true peak in the DFT spectrum (bottom). The middle and right panels illustrate the effect of increasing amounts of randomness incorporated into the sampling scheme on the spectral aliases. The incorporation of randomness can suppress artifacts in otherwise regular sampling schemes, such as radial sampling, as shown in Fig. 6.

Another way to increase the effective bandwidth is to sample from an oversampled grid. We saw earlier that oversampling can benefit uniform sampling approaches by increasing the dynamic range. When employed with NUS, oversampling has the effect of shifting sampling artifacts out of the original spectral window [47]. This effect is shown in Fig. 7.

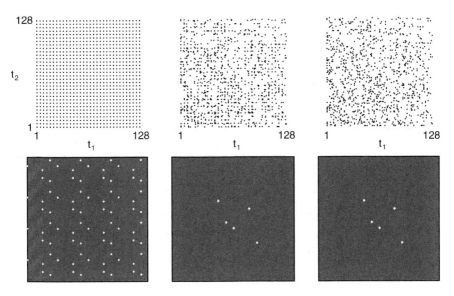

Fig. 5 MaxEnt spectra for synthetic two-dimensional data consisting of five exponentially decaying sinusoids plus noise. The *left-most panels* depict deliberate undersampling selecting every fourth point along both dimensions. The *center and right panels* depict blurred undersampling, RMS 0.625 and 1.25, respectively. *White* contour levels are plotted at multiples of 1.4 starting with 3% of the height of the highest peak

7 A Menagerie of Sampling Schemes

While the efficacy of a particular sampling scheme depends on a host of factors, including the nature of the signal being sampled, the PSF provides a useful first-order tool for comparing sampling schedules. Figure 8 illustrates examples of several common two-dimensional NUS schemes, together with PSFs computed for varying levels of coverage (30%, 10%, and 5%) of the underlying uniform grid. Some of the schemes are off-grid schemes, but they are approximated here by mapping onto a uniform grid. As noted previously, on-grid approximation of off-grid sampling schemes coupled with reconstruction methods such as MaxEnt gives results that are very similar to off-grid sampling. The PSF gives an indication of the distribution and magnitude of sampling artifacts for a given sampling scheme; schemes with PSFs that have very low values other than the central component give rise to weaker artifacts. Of course the PSF alone does not tell the whole story, because it does not address relative sensitivity. For example, while the random schedule has a PSF with very weak side-lobes, and gives rise to fewer artifacts than a radial sampling scheme for the same level of coverage, it has lower sensitivity for exponentially decaying sinusoids than a radial scheme (which concentrates more samples at short evolution times where the signal is strongest). Thus more than one metric is needed to assess the relative performance of different sampling schemes.

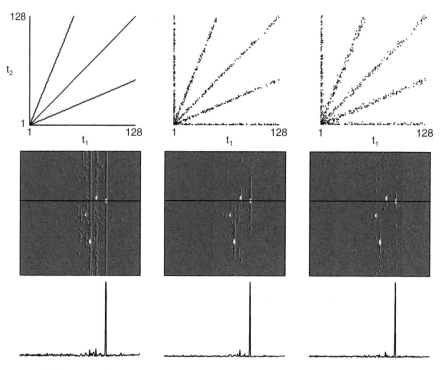

Fig. 6 ^{1}H/^{13}C plane (^{15}N chemical shift 121.96 ppm) from the HNCO spectrum of Ubiquitin, using data collected at 9.4 T (400 MHz for ^{1}H) on a Varian Inova instrument. Spectra were computed using MaxEnt reconstruction and radial sampling using five projections with different amounts of random "blurring" of the sampling schedule (RMS zero (none), 0.625 and 1.25, *left* to *right*). *Top*: sampling schedules. *Bottom*: MaxEnt spectra. Contour levels are chosen as in Fig. 5

7.1 Random and Biased Random Sampling

Exponentially-biased random sampling was the first general NUS approach applied to multidimensional NMR [22]. By analogy with matched filter apodization (which was first applied in NMR by Ernst, and maximizes the S/N of the uniformly-sampled DFT spectrum), Laue and colleagues reasoned that tailoring NUS so that the signal is sampled more frequently at short times, where the signal is strong, and less frequently when the signal is weak, would similarly improve S/N. They applied an exponential bias to match the decay rate of the signal envelope; we refer to this as envelope-matched sampling (EMS). Generalizations of the approach to sine-modulated signals, where the signal is small at the beginning, and constant-time experiments, where the signal envelope does not decay, were described by Schmieder et al. [32, 33].

Data Sampling in Multidimensional NMR: Fundamentals and Strategies

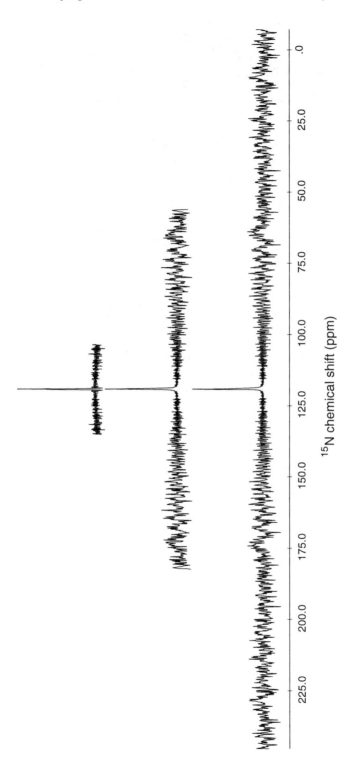

Fig. 7 Effect of oversampling. The *top* panel shows the two peaks and their associated sampling artifacts and the *middle* and *lower* panels show the same peaks using 4× and 8× SW. The sampling artifacts are shifted to extreme frequencies at the cost of line broadening

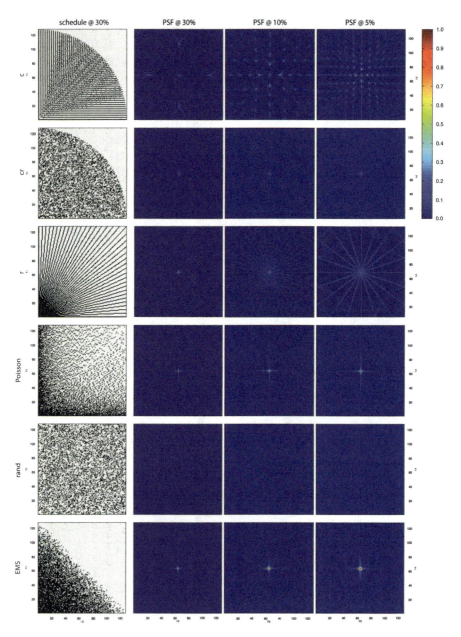

Fig. 8 (Continued)

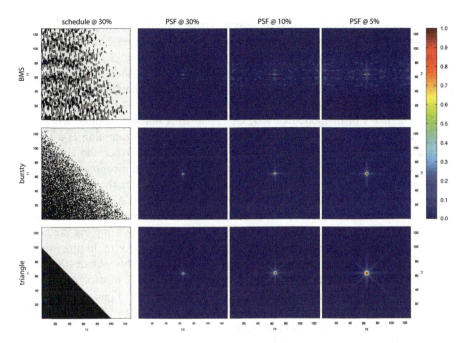

Fig. 8 A menagerie of sampling schemes. The first column depicts examples of two-dimensional sampling schemes that have been employed in NMR, for 30% coverage of a 128 × 128 uniform grid (i.e., approx. 4915 samples out of 16,384). Successive columns depict the PSF for 30%, 10%, and 5% coverage. The PSFs are normalized to the value of the central component, and the color coding is depicted on the far right. Sampling schedules depicted include (*c*) circular shell, (*cr*) randomized circular, (*r*) radial, (*Poisson*) Poisson gap, (*rand*) random, (*EMS*) envelope-matched, (*BMS*) beat-matched, (*burst*) bursty, and (*triangle*) triangular

7.2 Triangular

Somewhat analogous to the rationale behind exponentially-biased sampling, Delsuc and colleagues employed triangular sampling in two time dimensions to capture the strongest part of a two-dimensional signal [34]. The approach is easily generalized to arbitrary dimension.

7.3 Radial

Radial sampling results when the incrementation of evolution times is coupled, and is the approach employed by GFT, RD, and back-project reconstruction methods. Radial sampling has also found application in MRI. When a fully-dimensional spectrum is computed from a set of radial samples (e.g., BPR, radial FT, MaxEnt), the radial sampling vectors are typically chosen to somewhat uniformly span the orientations from 0° to 90°. When the fully-dimensional spectrum is not reconstructed, but instead the individual one-dimensional spectra (corresponding to projected cross sections through the fully-dimensional spectrum) are analyzed separately, the sampling angles are typically determined using a knowledge-based approach (HIFI, APSY [35, 36]). Prior knowledge about chemical shift distributions in proteins is employed to select sequentially radial vectors to minimize the likelihood of overlap in the projected cross section.

The successes of methods like RD, GFT, and BPR notwithstanding, when the aim is to reconstruct the fully-dimensional spectrum, radial sampling is a rather poor approach compared to less regular sampling schemes. When the aim is *not* to reconstruct the fully dimensional spectrum, but to analyze projections separately, a complete separate and dedicated infrastructure for the analysis is required (which comprises much of the effort behind GFT, HIFI, and APSY approaches). The advantage of reconstructing the fully dimensional spectrum is that the data are isomorphic with spectra computed using conventional uniform sampling methods, and the abundance of graphical and analysis tools that exist for multidimensional NMR data can be used to visualize and quantify the spectra. This includes XEASY [37], NMRDraw [38], NMRViewJ [39], Sparky [40], and a host of automated scripts for "strip" plots and sequential assignment of proteins. Figure 9 compares the use of radial sampling with exponentially biased random sampling in two indirect dimensions, using MaxEnt reconstruction to compute the 3D spectrum. The top panels depict contour plots using one, two, and three radial sampling vectors (from left to right). Below each panel are shown contour plots for spectra computed using biased random sampling using the same number of samples as the radial sampling example given directly above. The accuracy of the reconstruction of the 3D spectrum from a set of sparse samples is dramatically better when biased random sampling is used instead of radial sampling.

7.4 Concentric Rings

Coggins and Zhou introduced the concept of concentric ring sampling (CRS), and showed that radial sampling is a special case of CRS [29]. They showed that the DFT could be adapted to CRS (and radial sampling) by changing to polar coordinates from Cartesian coordinates (essentially by introducing the Jacobian for the coordinate transformation as weighting factors). Optimized CRS that linearly increases the number of samples in a ring as the radius increases and incorporates randomness were shown to provide resolution comparable to uniform sampling for the same measurement time, but with fewer sampling artifacts than radial sampling. They also showed that the discrete polar FT is equivalent to the result from weighted back projection reconstruction.

7.5 Spiral

Spiral sampling is used mainly in MRI, where it permits reduced exploration of k-space (and thus a reduction of scan time).

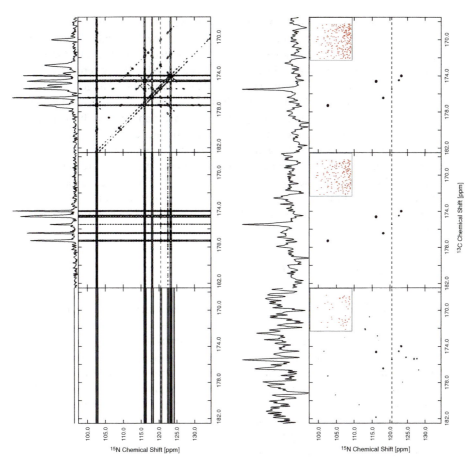

Fig. 9 HNCO spectra of ubiquitin. *Top panels* show the addition of 0°, 90°, and 30° projections of the two jointly sampled indirect dimensions at a proton chemical shift of 8.14 ppm, reconstructed using back projection reconstruction. Each projection contains 52 complex points; thus the total number of complex points sampled from *left* to *right* is 52, 104, and 156. The *lower panel* shows MaxEnt reconstruction using the same number of complex data points, distributed randomly along the nitrogen dimension (constant time) and with an exponentially decreasing sampling density decay rate corresponding to 15 Hz in the carbon dimension. A 1D trace at the position of the weakest peak present in the spectrum is shown at the top of each spectrum (indicated by a *dashed line*). The *insets* depict the sampling scheme

7.6 Beat-Matched Sampling

The concept of matching the sampling density to the signal envelope, in order to sample most frequently when the signal is strong and less frequently when it is weak, can be extended to match finer details of the signal. For example, a signal containing two strong frequency components will exhibit beats in the time domain signal separated by the reciprocal of the frequency difference between the components. As the signal becomes more complex, with more frequency components, more beats will occur corresponding to frequency differences between the various components. If one knows a priori the expected frequencies of the signal components, one can predict the location of the beats (and nulls, or zero-crossings), and tailor sampling accordingly. The procedure is entirely analogous to EMS, except that the sampling density is matched to the fine detail of predicted time-domain data, not just the signal envelope. We refer to this approach as beat-matched sampling (BMS). Possible applications where the frequencies are known a priori include relaxation experiments or multidimensional experiments in which scout scans or complementary experiments provide knowledge of the frequencies. In practice, BMS sampling schedules appear similar to EMS (e.g., exponentially biased) schedules; however, they tend not to be as robust, as small difference in noise level or small frequency shifts can have pronounced effects on the location of beats or nulls in the signal.

7.7 Poisson Gap Sampling

Hyberts and Wagner [41] noted empirically that the distribution of the gaps in a sampling schedule are also important. Long gaps near the beginning or end of a sample schedule were particularly detrimental. They adapted an idea employed in computer graphics, Poisson gap sampling, to generate sampling schedules that avoid long gaps while ensuring the samples are randomly distributed. Similar distributions can be generated using other approaches, for example quasi-random (e.g., Sobolev) sequences. In addition to being robust, Poisson gap sampling schedules show less variation with the random deviate than other sampling schemes. A potential weakness of Poisson gap sampling, however, is that the minimum distance between samples must not be too small, otherwise aliasing can become significant.

7.8 Burst Sampling

In burst or burst-mode sampling, short high-rate bursts are separated by stretches with no sampling. It effectively minimizes the *number* of large gaps, while ensuring that samples are spaced at the minimal spacing when sub-sampling from a grid. Burst sampling has found application in commercial spectrum analyzers and communications gear. In contrast to Poisson gap sampling, burst sampling ensures that most samples are separated by the grid spacing to suppress aliasing [42].

7.9 Nonuniform Averaging

The concept underlying EMS or BMS can be applied to the amount of signal averaging performed, in contexts where a significant number of transients are averaged to obtain sufficient sensitivity. In this sensitivity-limited regime, varying the number of transients in proportion to the signal envelope could be utilized in conjunction with uniform or nonuniform sampling in the time domain. An early application of this idea in NMR employed uniform sampling with nonuniform averaging, and computed the multidimensional DFT spectrum after first normalizing each FID by dividing by the number of transients summed at each indirect evolution time [43]. Although the results of this approach are qualitatively reasonable provided that the S/N is not too low, a flaw in the approach is that noise will not be properly weighted. A solution is to employ a method where appropriate statistical weights can be applied to each FID, e.g., MaxEnt or MLM reconstruction.

7.10 Random Phase Detection

We've seen how NUS artifacts are a manifestation of aliasing, and how randomization can mitigate the extent of aliasing. There is another context in which aliasing appears in NMR, and that is determining the *sign* of frequency components (i.e., the direction of rotation of the magnetization). An approach widely used in NMR to resolve this ambiguity is to detect simultaneously two orthogonal phases (simultaneous quadrature detection). When simultaneous quadrature detection is not feasible, for example in the indirect dimensions of a multidimensional experiment, oversampling by a factor of two together with placing the detector reference frequency outside the spectral window spanned by the signal can resolve the ambiguity (TPPI). Alternatively, two orthogonal phases can be detected sequentially (sequential quadrature detection). The total number of samples required to resolve the sign ambiguity is the same whether quadrature detection or oversampling is employed. Single-phase detection using uniform sampling with random phase (random phase detection, RPD) is able to resolve the frequency sign ambiguity without oversampling, as shown in Fig. 10. This results in a factor of two reduction in the number of samples required, compared to quadrature or TPPI detection methods, for each indirect dimension of a multidimensional experiment. For experiments not employing quadrature or TPPI detection, it provides a factor of two increase in resolution for each dimension.

7.11 Optimal Sampling?

Any sampling scheme, whether uniform or nonuniform, can be characterized by its effective bandwidth, dynamic range, resolution, sensitivity, and number of samples. Some of these metrics are closely related, and it is not possible to optimize all of them simultaneously. For example, minimizing the total number of samples (and thus the experiment time) invariably increases the magnitude of sampling artifacts. Furthermore, a sampling scheme that is optimal for one signal will not necessarily

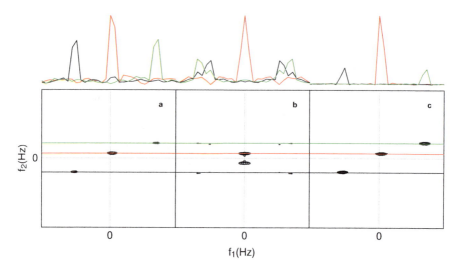

Fig. 10 Two-dimensional f_1/f_2 cross-sections from four-dimensional N,C-NOESY data for the DH1 domain of Kalirin. One dimensional cross sections parallel to the f_1 axis at the f_2 frequencies indicated by the *colored lines* are shown above each panel. *Panel A* is the real/real component of the two dimensional DFT spectrum using quadrature detection in all dimensions. *Panel B* is the DFT spectrum obtained using only the real/real/real component from the three indirect time dimensions of the time domain data. *Panel C* is the maximum entropy spectrum obtained using random phase detection. *Panels B* and *C* employ 1/8th the number of samples used in *panel A*

be optimal for a signal containing frequency components with different characteristics. Thus the design of efficient sampling schemes involves tradeoffs. Simply put, no single NUS scheme will be best suited for all experiments.

8 Concluding Remarks

The use of NUS in all its guises is transforming the practice of multidimensional NMR, most importantly by lifting the sampling limited obstacle to obtaining the potential resolution in indirect dimensions afforded by ultra high-field magnets. NUS is also beginning to have tremendous impact in MRI, where even small reductions in the time required to collect an image can have tremendous clinical impact. For all the successes using NUS, our understanding of how to design optimal sampling schemes remains incomplete. A major limitation is that we lack a comprehensive theory able to predict the performance of a given NUS scheme a priori. This in turn is related to the absence of a consensus on performance metrics, i.e., measures of spectral quality. Ask any three NMR spectroscopists to quantify the quality of a spectrum and you are likely to get three different answers. Further advances in NUS will be enabled by the development of robust, shared metrics. An additional hurdle has been the absence of a common set of test or reference data, which is necessary for critical comparison of competing approaches. Once shared metrics and reference data are established, we anticipate rapid additional improvements in the design and application of NUS to multidimensional NMR spectroscopy.

Acknowledgements We thank Gerhard Wagner for providing a pre-publication manuscript for the contribution by Hyberts and Wagner in this volume. We thank Sven Hyberts for providing the Poisson gap sampling schedules used in Fig. 8, and for helpful discussions. JCH gratefully acknowledges support from the US National Institutes of Health (grants GM047467 and RR020125).

References

1. Ernst RR, Anderson WA (1966) Application of Fourier transform spectroscopy to magnetic resonance. Rev Sci Instrum 37:93–102
2. Hoch JC, Stern AS (1996) NMR data processing. Wiley-Liss, New York
3. Beckman RA, Zuiderweg ERP (1995) Guidelines for the use of oversampling in protein NMR. J Magn Reson A113:223–231
4. Delsuc MA, Lallemand JY (1986) Improvement of dynamic range in NMR by oversampling. J Magn Reson 69:504–507
5. Rovnyak D, Hoch JC, Stern AS, Wagner G (2004) Resolution and sensitivity of high field nuclear magnetic resonance spectroscopy. J Biomol NMR 30:1–10
6. Jeener J (1971) Oral presentation, Ampere International Summer School, Yugoslavia
7. States DJ, Haberkorn RA, Ruben DJ (1982) A two-dimensional nuclear Overhauser experiment with pure absorption phase in four quadrants. J Magn Reson 48:286–292
8. Stern AS, Li K-B, Hoch JC (2002) Modern spectrum analysis in multidimensional NMR spectroscopy: comparison of linear-prediction extrapolation and maximum-entropy reconstruction. J Am Chem Soc 124:1982–1993
9. Wernecke SJ, D' Addario LR (1977) Maximum entropy image reconstruction. IEEE Trans Comput 26:351–364
10. Skilling J, Bryan R (1984) Maximum entropy image reconstruction: general algorithm. Mon Not R Astron Soc 211:111–124
11. Stern AS, Donoho DL, Hoch JC (2007) NMR data processing using iterative thresholding and minimum l(1)-norm reconstruction. J Magn Reson 188:295–300
12. Hyberts SG, Heffron GJ, Tarragona NG, Solanky K, Edmonds KA, Luithardt H, Fejzo J, Chorev M, Aktas H, Colson K, Falchuk KH, Halperin JA, Wagner G (2007) Ultrahigh-resolution (1)H-(13)C HSQC spectra of metabolite mixtures using nonlinear sampling and forward maximum entropy reconstruction. J Am Chem Soc 129:5108–5116
13. Bretthorst GL (1990) Bayesian Analysis I. Parameter estimation using quadrature NMR models. J Magn Reson 88:533–551
14. Chylla RA, Markley JL (1993) Improved frequency resolution in multidimensional constant-time experiments by multidimensional Bayesian analysis. J Biomol NMR 3:515–533
15. Chylla RA, Markley JL (1995) Theory and application of the maximum likelihood principle to NMR parameter estimation of multidimensional NMR data. J Biomol NMR 5:245–258
16. Jaravine V, Ibraghimov I, Orekhov VY (2006) Removal of a time barrier for high-resolution multidimensional NMR spectroscopy. Nat Methods 3:605–607
17. Bro R (1997) PARAFAC. Tutorial and applications. Chemometr Intell Lab Syst 38:149–171
18. Kazimierczuk K, Kozminski W, Zhukov I (2006) Two-dimensional Fourier transform of arbitrarily sampled NMR data sets. J Magn Reson 179:323–328
19. Press WH, Flannery BP, Teukolsky SA, Vetterling WT (1992) Numerical recipes in Fortran. Cambridge University Press, Cambridge

20. Pannetier N, Houben K, Blanchard L, Marion D (2007) Optimized 3D-NMR sampling for resonance assignment of partially unfolded proteins. J Magn Reson 186:142–149
21. Bodenhausen G, Ernst RR (1969) The accordion experiment, a simple approach to three-dimensional NMR spectroscopy. J Magn Reson 45(1981):367–373
22. Barna JCJ, Laue ED, Mayger MR, Skilling J, Worrall SJP (1987) Exponential sampling: an alternative method for sampling in two dimensional NMR experiments. J Magn Reson 73:69
23. Carr PA, Fearing DA, Palmer AG (1998) 3D accordion spectroscopy for measuring15N and13CO relaxation rates in poorly resolved NMR spectra. J Magn Reson 132:25–33
24. Chen K, Tjandra N (2009) Direct measurements of protein backbone 15 N spin relaxation rates from peak line-width using a fully-relaxed Accordion 3D HNCO experiment. J Magn Reson 197:71–76
25. Szyperski T, Wider G, Bushweller JH, Wüthrich K (1993) Reduced dimensionality in triple resonance NMR experiments. J Am Chem Soc 115:9307–9308
26. Szyperski T, Wider G, Bushweller JH, Wüthrich K (1993) 3D 13 C-15 N-heteronuclear two-spin coherence spectroscopy for polypeptide backbone assignments in 13 C-15 N-double-labeled proteins. J Biomol NMR 3:127–132
27. Kim S, Szyperski T (2003) GFT NMR, a new approach to rapidly obtain precise high-dimensional NMR spectral information. J Am Chem Soc 125:1385–1393
28. Coggins BE, Zhou P (2006) Polar Fourier transforms of radially sampled NMR data. J Magn Reson 182:84–95
29. Coggins BE, Zhou P (2007) Sampling of the NMR time domain along concentric rings. J Magn Reson 184:207–221
30. Bretthorst GL (2001) Nonuniform sampling: bandwidth and aliasing. In: Rychert J, Erickson G, Smith CR (eds) Maximum entropy and Bayesian methods in science and engineering. Springer, New York, pp 1–28
31. Bretthorst GL (2008) Nonuniform sampling: bandwidth and aliasing. Concepts Magn Reson 32A:417–435
32. Schmieder P, Stern AS, Wagner G, Hoch JC (1993) Application of nonlinear sampling schemes to COSY-type spectra. J Biomol NMR 3:569–576
33. Schmieder P, Stern AS, Wagner G, Hoch JC (1994) Improved resolution in triple-resonance spectra by nonlinear sampling in the constant-time domain. J Biomol NMR 4:483–490
34. Aggarwal K, Delsuc MA (1997) Triangular sampling of multidimensional NMR data sets. Magn Reson Chem 35:593–596
35. Eghbalnia HR, Bahrami A, Tonelli M, Hallenga K, Markley JL (2005) High-resolution iterative frequency identification for NMR as a general strategy for multidimensional data collection. J Am Chem Soc 127:12528–12536
36. Hiller S, Fiorito F, Wüthrich K (2005) Automated projection spectroscopy (APSY). Proc Natl Acad Sci USA 102:10876–10888
37. Bartels C, Xia T-H, Billeter M, Güntert P, Wüthrich K (1995) The program XEASY for computer-supported NMR spectral analysis of biological macromolecules. J Biomol NMR 5:1–10
38. Delaglio F, Grzesiek S, Vuister GW, Zhu G, Pfeifer J, Bax A (1995) NMRPipe: a multidimensional spectral processing system based on UNIX pipes. J Biomol NMR 6:277–293
39. Johnson BA (2004) Using NMR view to visualize and analyze the NMR spectra of macromolecules. Methods Mol Biol 278:313–352
40. Goddard TD, Kneller DG (2006) SPARKY 3, University of California, San Francisco
41. Hyberts SG, Takeuchi K, Wagner G (2010) Poisson-gap sampling and forward maximum entropy reconstruction for enhancing the resolution and sensitivity of protein NMR data. J Am Chem Soc 132:2145–2147
42. Maciejewski MW, Qui HZ, Rujan I, Mobli M, Hoch JC (2009) Nonuniform sampling and spectral aliasing. J Magn Reson 199:88–93
43. Kumar A, Brown SC, Donlan ME, Meier BU, Jeffs PW (1991) Optimization of two-dimensional NMR by matched accumulation. J Magn Reson 95:1–9

44. Mehdi M, Alan SS, Jeffrey CH (2006) Spectral Reconstruction Methods in Fast NMR: Reduced Dimensionality, Random Sampling, and Maximum Entropy Reconstruction. J Magn Reson 192:96–105
45. Mehdi M, Alan SS, Wolfgang B, Glenn FK, Jeffrey CH, (2010) A non-uniformly sampled 4D HCC(CO)NH-TOCSY experiment processed using maximum entropy for rapid protein sidechain assignment. J Magn Reson 204:160–164
46. Hoch JC, Maciejewski MW, Filipovic B (2008) Randomization improves sparse sampling in multidimensional NMR. J Magn Reson 193:317–20
47. Mark WM, Harry Z, Qui IR, Mehdi M, Jeffrey CH (2009) Nonuniform sampling and spectral aliasing. J Magn Reson 199:88–93

Generalized Fourier Transform for Non-Uniform Sampled Data

Krzysztof Kazimierczuk, Maria Misiak, Jan Stanek,
Anna Zawadzka-Kazimierczuk, and Wiktor Koźmiński

Abstract Fourier transform can be effectively used for processing of sparsely sampled multidimensional data sets. It provides the possibility to acquire NMR spectra of ultra-high dimensionality and/or resolution which allow easy resonance assignment and precise determination of spectral parameters, e.g., coupling constants. In this chapter, the development and applications of non-uniform Fourier transform is presented.

Keywords Biomolecular NMR · Multidimensional NMR · Non-linear sampling · Sparse sampling

Contents

1 Introduction 80
2 Fourier Transform: Basics 83
    2.1 Definition 83
    2.2 Multidimensional FT 85
    2.3 FT: Two Basic Features 86
3 Fourier Transform of the NMR Signal 88
    3.1 Perfect FID 88
    3.2 Measured FID 89
4 Non-Uniform Sampling Schemes 95
    4.1 Radial Sampling 95
    4.2 Concentric Rings Sampling 96
    4.3 Spiral Sampling 96

K. Kazimierczuk
Faculty of Chemistry, University of Warsaw, Pasteura 1, 02093 Warsaw, Poland
Swedish NMR Centre, University of Gothenburg, Box 465, 405 30 Gothenburg, Sweden

M. Misiak, J. Stanek, A. Zawadzka-Kazimierczuk, and W. Koźmiński (✉)
Faculty of Chemistry, University of Warsaw, Pasteura 1, 02093 Warsaw, Poland
e-mail: kozmin@chem.uw.edu.pl

	4.4	Random Sampling	96
	4.5	Weighted Samples and Weighted Probability	97
5	Methods of Integration		97
6	Sparse Sampling and FT as a Linear Algebra Issue		98
7	Suppression of Sampling Artifacts in FT Spectra		101
	7.1	The Principle of CLEAN Algorithm	101
	7.2	Development of the CLEAN Algorithm	104
	7.3	The Early Applications of CLEAN to NMR Spectroscopy	105
	7.4	Algorithms Related to CLEAN	107
8	FT as a Tool for Large Evolution Time Domain		108
	8.1	Features of "Sampling Noise"	108
	8.2	Sparse MFT	110
9	Applications		113
	9.1	Development and Implementations	113
	9.2	Easy Resonance Assignment in Proteins Using the Spectra of High Dimensionality	115
	9.3	Determination of Coupling Constants in Proteins	115
	9.4	Heteronuclear-Edited NOESY Experiments	118
	9.5	3D Spectra of Complex Organic Compounds	119
10	Conclusions		122
References			122

1 Introduction

Nuclear Magnetic Resonance (NMR) spectroscopy is one of the most important tools in structural studies of chemical compounds, ranging from small molecules up to medium-sized proteins. NMR spectra provide valuable information about molecular structure, interactions and dynamics. However, there is still a need for more robust and more effective methods of acquisition and processing of NMR data.

Early NMR spectroscopy utilized the Continuous Wave (CW) detection technique. It was based on continuous sweeping of the B_0 field strength, or the frequency of electromagnetic wave, through the resonance conditions of nuclei in the assumed spectral range. The main drawbacks of CW detection were low sensitivity and loss of time needed for sweeping through empty spectral regions. The breakthrough in NMR spectroscopy was the development of pulse excitation for generation of the Free Induction Decay (FID) signal, and the observation that the time dependent FID signal and the NMR spectrum can be converted one into the other by applying the Fourier transform (FT) [1]. This method greatly shortened spectral acquisition times and enabled the development of thousands of pulse sequences for numerous emerging applications. New experimental methods allowed the more accurate determination of parameters which were earlier difficult to obtain or inaccessible.

Despite considerable progress in the field, NMR spectroscopy still has two significant limitations: the intrinsically low sensitivity, due to the low Boltzmann polarization of nuclear spins in thermal equilibrium, and the low dispersion of observed frequencies, due to small differences in nuclear shielding by surrounding electrons for nuclei of the same kind. The first problem is continuously circumvented by technological developments, i.e., construction of higher field magnets, cryogenically cooled probe-heads and pre-amplifiers, modern electronics, cleaner RF sources, and recently, introduction of the Dynamic Nuclear Polarization (DNP) technique enabling sensitivity gain of even two orders of magnitude. The new generations of NMR spectrometers feature higher sensitivity and allow studies of large molecules at lower concentration. The problem of resolving overlapped resonances is more severe. Even spectra of simple molecules often exhibit a peak overlap. Additionally, the assignment of signal frequencies to the respective nuclei could be difficult and sometimes impossible. To some extent, in simple cases, the problem can be solved by employing a stronger magnetic field, but the general approach to resolve the resonances is to spread them in different frequency dimensions of multidimensional spectra. The idea was practically implemented by the indirect sampling of spins evolution and was referred to as two-dimensional NMR spectroscopy [2, 3]. This development not only allowed resolution of individual peaks by introducing additional spectral dimensions, but also facilitated spectral assignment by detecting groups of mutually interacting nuclei which give rise to correlation peaks. The application of multiple polarization transfer revealed other important aspects of multidimensional spectroscopy: sensitivity enhancement by excitation and observation of FID signal of sensitive, high-γ spins, and observation of directly undetectable multiple quantum coherences. At the beginning, the two-dimensional NMR techniques were demonstrated to be useful for examination of small organic molecules. Soon, homonuclear 2D NMR experiments were successfully applied for studies of biological macromolecules in solution [4]. Later, with increasing availability of isotopically enriched proteins, significant improvement was achieved by introduction of triple-resonance three- and four-dimensional experiments utilizing scalar couplings for polarization transfers [5–9]. However, due to the reasons given below, acquisition of multidimensional NMR spectra with a sufficient resolution in all frequency dimensions can be an extremely time consuming task.

Indirect sampling of spin coherences evolution, the key concept of multidimensional NMR experiments, is realized in a parametric way. This means that to sample a point of indirect time space, a specific delay (or delays) in a pulse sequence should be set to achieve desired evolution time, and then one directly observed FID signal is acquired. As a consequence, in order to acquire a multidimensional spectrum one needs to record many single FID signals. The overall measurement time grows rapidly with a number of indirectly sampled dimensions and a desired resolution. A conventional N-dimensional experiment requires acquisition of $2^{N-1} \cdot k_1 \cdot k_2 \cdot \ldots \cdot k_{N-1}$ single FID signals (where $k_i = sw_i \cdot t_{\max i}$, is the number of points in the ith dimension, sw_i and $t_{\max\,i}$ are the required spectral width and maximum evolution time respectively, and 2^{N-1} is the number of components needed for quadrature detection). Conventional sampling is performed with points placed on a Cartesian grid. In each dimension, spacing between points is related to the expected range of frequencies by the Nyquist Theorem (see Sect. 3.2.3). Thus, fulfilling the Nyquist theorem implicitly limits the maximum evolution time and, therefore, the obtainable resolution for the given duration of experiment. In a case of directly detected dimension the above limitation is insignificant. Here, data points are successively sampled by conversion of a voltage in a receiver circuit into numbers reflecting signal amplitude. The acquisition of a whole signal has to be performed in one step, and it usually takes from a fraction of second to a few seconds until the signal decays below the noise level. This does not significantly extend the experiment duration. Thus, the best possible resolution is almost always achievable at no additional cost. Moreover, in modern NMR spectrometers, oversampling is usually employed, i.e., more points than necessary are sampled in order to improve spectral dynamic range and enable digital filtering [10].

Sampling limitations have more severe consequences in the case of indirectly sampled dimensions, where acquisition of each sampling point takes up to a few seconds. Even in 3D NMR experiments of proteins, featuring relatively fast transverse relaxation, it is almost impossible to reach the natural (determined by relaxation) line width in a reasonable experimental time. Limited experiment duration causes signal truncation and results in broadened spectral peaks, according to the Fourier Uncertainty Principle [11].

The problem of sampling requirements in multidimensional NMR is becoming relatively more severe with increasing B_0 fields. The stronger magnetic field increases proportionally separation between resonances; however, at the same time, it broadens spectral regions of interest. Hence, an x-fold increase in B_0 causes the necessity of x^{N-1}-fold extension of time required for N-dimensional experiment in order to preserve the peak width. This effect, although usually of minor importance for 2D experiments, became significant for the larger number of dimensions.

In the last 10 years many approaches were proposed to overcome the sampling limitation problem. The most straightforward is the modification of pulse sequences to allow increased repetition rate of FID signal acquisition, which leads to reduction of the experiment time [12–14]. It was also shown that the spatial encoding of spectral frequencies can be employed for measurement of multidimensional spectra in a single scan [15–20]. However, most of the efforts to accelerate acquisition of multidimensional NMR spectra were dedicated to the reconstruction of so-called "sparsely sampled" spectra, i.e., with less data points than required by Nyquist condition. Both experiment duration and desired resolution can be optimized by the use of sparse sampling. The simplest version of sparse sampling is a straightforward signal truncation. In such a case it is possible to attempt signal extrapolation using linear prediction [21] or filter diagonalization methods [22–24]. Enhanced spectral resolution can also be achieved from relatively highly truncated data sets employing Covariance Spectroscopy [25–30], and some variants of the maximum entropy method [31]. Another simple approach to undersampling is increasing of the distance between points which leads to shortened experiment duration, at the expense of peak aliasing. Thus, if chemical shifts are known from other experiments, assignment of cross-peaks is still possible [32]. The sparse sampling can also be applied in order to extend the sampled space in several ways. Among them, two are of particular importance: sampling at constant intervals, but along the radius in a time domain [33], or randomly [34]. The former option is utilized in projection spectroscopy, and requires the algebraic decoding of peak frequencies

[35–39], or the reconstruction of multidimensional spectrum [40–43]. The latter enables one to reconstruct a fully-dimensional spectrum featuring improved resolution which is acquired faster than conventionally. The sparsely and randomly sampled data sets can be processed using FT [44–46], maximum entropy [47–49] or multidimensional decomposition [50–52] methods.

In this review we will focus on applications of FT to processing of non-uniformly (sparsely) data sets devoted to the reconstruction of high-resolution multidimensional NMR spectra.

2 Fourier Transform: Basics

2.1 Definition

FT is a mathematical operation that converts function $s(t)$ into function $S(f)$ according to the formula

$$S(f) = \int_{-\infty}^{+\infty} s(t) \cdot e^{-2\pi i f t} dt, \qquad (1)$$

which, for convenience, may be denoted as a linear operator "FT" acting on $s(t)$:

$$S(f) = \text{FT}[s(t)]. \qquad (2)$$

Function $e^{-2\pi i f t}$ that multiplies signal $s(t)$ is often referred to as a *transform kernel*.

Both t and f are real variables, while $s(t)$ and $S(f)$ may be complex in general. In many fields of signal processing (including NMR spectroscopy), the two variables correspond to *time* and *frequency* domains. Function $s(t)$ is a time-domain signal recorded in the experiment. Function $S(f)$ is its *frequency representation*, i.e., it shows how a signal can be decomposed into oscillatory functions of frequencies f. Knowing frequency representation of a signal, one can retrieve $s(t)$ by applying Inverse Fourier Transform (IFT):

$$s(t) = \int_{-\infty}^{+\infty} S(f) \cdot e^{2\pi i f t} df. \qquad (3)$$

Hence, $s(t)$ and $S(f)$ are equivalent representations of a signal and are often referred to as a *Fourier pair*. For the simplest infinite oscillatory signal of frequency v the Fourier pair is

$$e^{2\pi i vt} \xrightleftharpoons[\text{IFT}]{\text{FT}} \delta(f-v), \tag{4}$$

where $\delta(f-v)$ is the Dirac delta, and can be informally thought of as an infinitely narrow and infinitely high peak centered at v:

$$\delta(f-v) = \begin{cases} +\infty & \text{for } f = v \\ 0 & \text{for } f \neq v \end{cases}. \tag{5}$$

The result can be explained by the orthogonality of oscillatory exponentials that are *basis functions* for FT, i.e.,

$$\int_{-\infty}^{+\infty} e^{2\pi i vt} \cdot e^{-2\pi i ft} dt = \begin{cases} +\infty & \text{for } f = v \\ 0 & \text{for } f \neq v \end{cases}. \tag{6}$$

Thus, as the result of FT, one obtains a function that reaches high values for coordinates corresponding to frequencies present in a signal. This function, called *spectrum*, is of particular interest, especially in scientific tasks. Representing oscillatory time-domain signal as a peak in frequency domain often provides better insight into physical phenomena, as discussed in the next section.

Description of measured signals based on complex numbers may be quite confusing and requires brief explanation. Notably, complex signal is artificially constructed from actually measured real-valued signals of the same frequencies and amplitudes, but shifted in phase by $\frac{\pi}{2}$:

$$s(t) = s_{\cos}(t) + i s_{\sin}(t), \tag{7}$$

e.g., $s(t) = e^{2\pi i vt}$ consists of $s_{\cos}(t) = \cos(2\pi vt)$, and $s_{\sin}(t) = \sin(2\pi vt)$.

Equivalently, one may use a two-dimensional vector to describe the complex signal:

$$s(t) = \begin{pmatrix} s_{\cos}(t) \\ s_{\sin}(t) \end{pmatrix}. \tag{8}$$

This notation will be used in discussion of multidimensional FT in the next section. Variants of FT featuring real-valued kernel, i.e., *Cosine FT* (Cos-FT) and *Sine FT* (Sin-FT), can be defined as

$$S_{\cos}(f) = \int_{-\infty}^{+\infty} s(t) \cdot \cos(2\pi ft) dt, \tag{9a}$$

$$S_{\sin}(f) = \int_{-\infty}^{+\infty} s(t) \cdot \sin(2\pi ft) dt. \tag{9b}$$

Again, for convenience, operator notation can be introduced:

$$S_{\cos}(f) = \mathrm{FT}_{\cos}[s(t)], \tag{10a}$$

$$S_{\sin}(f) = \mathrm{FT}_{\sin}[s(t)]. \tag{10b}$$

Complex FT of a complex signal can be described as a sum of *Cosine FT* and *Sine FT*:

$$S(f) = \mathrm{FT}_{\cos}[s_{\cos}(t)] + \mathrm{FT}_{\sin}[s_{\sin}(t)] - i(\mathrm{FT}_{\sin}[s_{\cos}(t)] - \mathrm{FT}_{\cos}[s_{\sin}(t)]). \tag{11}$$

This notation allows easy visualization of the essence of complex FT (see Fig. 1).

2.2 Multidimensional FT

Fourier Transform can be extended to N dimensions:

$$S(\vec{f}) = \int_{\mathfrak{R}^N} s(\vec{t}) \cdot \left[\begin{pmatrix} \cos(2\pi f_1 t_1) \\ \sin(2\pi f_1 t_1) \end{pmatrix} \otimes \begin{pmatrix} \cos(2\pi f_2 t_2) \\ \sin(2\pi f_2 t_2) \end{pmatrix} \otimes \ldots \otimes \begin{pmatrix} \cos(2\pi f_N t_N) \\ \sin(2\pi f_N t_N) \end{pmatrix} \right] d\vec{t}, \tag{12}$$

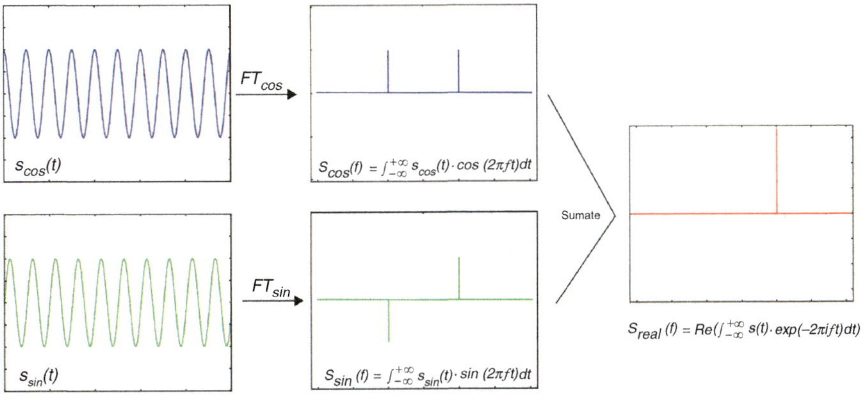

Fig. 1 Idea of complex FT. Two signals of the same frequency and amplitude, shifted in phase by $\pi/2$, $s_{\cos}(t)$ and $s_{\sin}(t)$ are transformed with cosine and sine FT and added. This may be described as one complex operation on one complex signal $s(t) = s_{\cos}(t) + i s_{\sin}(t)$

where $\vec{f}, \vec{t}$, are N-dimensional vectors:

$$\vec{f} = (f_1, f_2, ..., f_N),$$

$$\vec{t} = (t_1, t_2, ..., t_N).$$

It is noteworthy, that the transform kernel is represented by direct product ($\otimes$) of one-dimensional complex functions. The kernel is thus a 2^N-dimensional vector. One can represent both signal $\hat{s}(\vec{t})$ and spectrum $\hat{S}(\vec{f})$ in a similar fashion:

$$\hat{s}(\vec{t}) = s(t_1) \otimes s(t_2) \otimes ... \otimes s(t_N),$$

$$\hat{S}(\vec{f}) = S(f_1) \otimes S(f_2) \otimes ... \otimes S(f_N),$$

where $S(f_i) = \mathrm{FT}[s(t_i)]$.

The first element of $\hat{S}(\vec{f})$ corresponds to the real part of a spectrum.

The FT of the simplest multidimensional signal is thus a multidimensional delta function, i.e.,

$$\begin{pmatrix} \cos(2\pi v_1 t_1) \\ \sin(2\pi v_1 t_1) \end{pmatrix} \otimes \begin{pmatrix} \cos(2\pi v_2 t_2) \\ \sin(2\pi v_2 t_2) \end{pmatrix} \otimes ... \otimes \begin{pmatrix} \cos(2\pi v_N t_N) \\ \sin(2\pi v_N t_N) \end{pmatrix} \xrightarrow[\mathrm{IFT}]{\mathrm{FT}} \delta(f_1 - v_1, f_2 - v_2, ..., f_N - v_N), \quad (13)$$

where $\delta(f_1 - v_1, f_2 - v_2, ..., f_N - v_N) = \delta(f_1 - v_1) \otimes \delta(f_2 - v_2) ... \otimes \delta(f_N - v_N)$.

Again, signal frequency (frequencies) is (are) clearly visualized in the spectral domain as a "peak" centered at $(v_1, v_2, ..., v_N)$. Its position informs about correlated frequencies present in the multidimensional signal, which is usually the most essential experimental information.

2.3 FT: Two Basic Features

At the end of this section, we would like to mention two of the most important features of FT, which will be helpful in the analysis of specific case of NMR signal. These are:

1. Linearity:

$$\mathrm{FT}[\alpha f(t) + \beta g(t)] = \alpha \mathrm{FT}[f(t)] + \beta \mathrm{FT}[g(t)]. \quad (14)$$

2. Convolution Theorem:

$$FT[f(t) \cdot g(t)] = FT[f(t)] * FT[g(t)], \tag{15}$$

where $*$ denotes convolution, defined as (see also Fig. 2a, b)

$$u(x) * v(x) = \int_{-\infty}^{\infty} u(x) \cdot v(y - x) dy. \tag{16}$$

The two above features of FT will help us to evaluate how simple manipulations of the signal, like multiplication and addition, affect its spectrum. Notably, using only these two kinds of operations allows one to change from the monochromatic, non-decaying, perfectly continuous and infinite signal discussed above to the actually measured NMR signal.

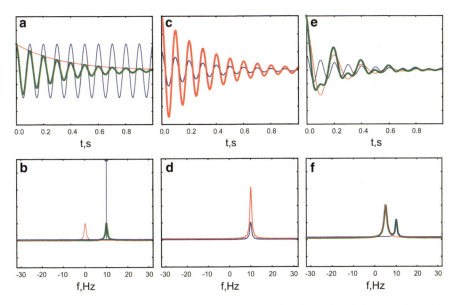

Fig. 2 The main features of FID signal and its spectrum: (**a**) Relaxing NMR signal (*bold line*) is product of decaying exponential function and oscillatory function (*thin lines*). (**b**) Spectrum of relaxing NMR signal (*bold line*) is convolution of Lorentzian function and a delta peak (*thin lines*). (**c**) Relaxing NMR signal of some amplitude A (*bold line*) is decaying sinusoid (*thin line*) multiplied by constant A. (**d**) Spectrum of relaxing NMR signal of some amplitude A (*bold line*) is Lorentzian peak (*thin line*) multiplied by constant A. (**e**) Multi-component NMR signal (*bold line*) is a sum of decaying components of different amplitudes (*thin lines*). (**f**) Spectrum of multi-component NMR signal (*bold line*) is a sum of spectra of individual components (*thin lines*)

3 Fourier Transform of the NMR Signal

3.1 Perfect FID

FID signal is a time-domain function resulting from NMR measurement. Although it is quite complicated, one can easily separate its features and discuss how they manifest themselves in frequency representation, i.e., a spectrum. These features are: relaxation, signal amplitude and multiple components (see Fig. 2).

3.1.1 Relaxation

NMR signal decays exponentially with time (or with "times", in multidimensional cases). This can be represented by element-wise multiplication of a signal $\hat{s}_0(\vec{t}) = \left(e^{-i2\pi v_1 t_1}, e^{-i2\pi v_2 t_2}, \ldots e^{-i2\pi v_N t_N}\right)^T$ by decaying exponential:

$$\hat{s}_1(\vec{t}) = \hat{s}_0(\vec{t}) \left(e^{-R_1 t_1}, e^{-R_2 t_2}, \ldots e^{-R_N t_N}\right)^T. \tag{17}$$

The FT of $\hat{s}_1(\vec{t})$ is, according to statements from Sect. 2.3, a convolution of multidimensional peak $\delta(f_1 - v_1, f_2 - v_2, \ldots, f_N - v_N)$ (i.e., $\text{FT}[\hat{s}_0(\vec{t})]$) and Lorentzian function (being FT of a decay function):

$$\text{FT}[\hat{s}_1(\vec{t})] = \delta(f_1 - v_1, f_2 - v_2, \ldots, f_N - v_N)$$
$$* \left(\frac{1}{R_1 + if_1}, \frac{1}{R_2 + if_2}, \ldots \frac{1}{R_N + if_N}\right). \tag{18}$$

The result is Lorentzian peak centered at $(v_1, v_2, \ldots, v_N)$.

3.1.2 Amplitude

NMR signal has certain amplitude. This can be represented by multiplying $\hat{s}_1(\vec{t})$ by some constant A:

$$\hat{s}_2(\vec{t}) = A \cdot \hat{s}_1(\vec{t}). \tag{19}$$

Obviously, multiplying the signal by a constant is equivalent to multiplication of the spectrum by the same constant:

$$\text{FT}[A \cdot \hat{s}_1(\vec{t})] = A \cdot \text{FT}[\hat{s}_1(\vec{t})]. \tag{20}$$

3.1.3 Multiple Components

An NMR signal consists of multiple components, corresponding to groups of equivalent spin systems. Each of the components has its own amplitude and relaxation parameters:

$$\hat{s}_3(\vec{t}) = \sum_i \hat{s}_2^i(\vec{t}). \qquad (21)$$

To summarize, as a model of multidimensional FID signal, one can use an oscillatory function consisting of multiple, decaying components of various amplitudes. The spectrum of such a signal is built of Lorentzian peaks centered at frequency coordinates corresponding to component frequencies. Peak heights are proportional to component amplitudes in time domain. Peak half-widths are the inverse of signal decay rates.

3.2 Measured FID

The above model gives an idea of what a *perfect* signal and its spectrum look like.

The *real* output of an NMR experiment is quite far from the model. Three factors are most important here, namely noise, finite measurement time, and sampling.

3.2.1 Noise

NMR signal contains some random noise $\varepsilon(t)$. Its spectrum is thus the sum of two FTs:

$$\hat{S}(\vec{f}) = \mathrm{FT}[\hat{s}_3(\vec{t})] + \mathrm{FT}[\varepsilon(\vec{t})]. \qquad (22)$$

Assuming that the noise is *white* and *Gaussian*, i.e., its amplitude is independent of frequency and described by Gaussian distribution, the signal to noise ratio is proportional to the square root of the number of measurements (see Fig. 3).

3.2.2 Finite Measurement Time

Obviously, maximum time of spin evolution cannot be infinite (which would be pointless anyway, because of relaxation). This limit can be represented by multiplying signal by step function $\Pi(t_1, t_2..., t_m)$:

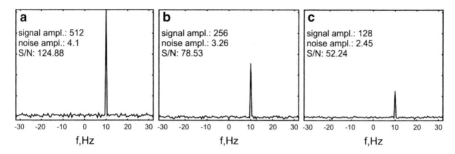

Fig. 3 Peak amplitude, noise level, and signal-to-noise ratio for spectrum of non-decaying signal of frequency 10 Hz, sampled with: (**a**) 512, (**b**) 256, (**c**) 128 pts. Noise is white and Gaussian

$$\Pi(t_1, t_2...t_m) = \begin{cases} 1 & t_i < t_{i\max}, i = 1, ..., m \\ 0 & \text{otherwise} \end{cases}, \quad (23)$$

where $t_{i\max}$ is a maximum evolution time set in the ith spectral dimension. Signal $\hat{s}_{t\max}(\vec{t})$, i.e., time-limited noiseless NMR signal, can be described as

$$\hat{s}_{t\max}(\vec{t}) = \hat{s}_3(\vec{t}) \cdot \Pi(t_1, t_2..., t_m). \quad (24)$$

The effect of multiplication in time domain is, according to the Convolution Theorem, the convolution in frequency domain:

$$\text{FT}[\hat{s}_{t\max}(\vec{t})] = \text{FT}[\hat{s}_3(\vec{t})] * \text{FT}[\Pi(t_1, t_2..., t_m)]. \quad (25)$$

The FT of step function is a *sinc* function of width inversely proportional to $t_{i\max}$ (see Fig. 4):

$$\text{FT}\{\Pi(\vec{t})\} = \frac{\sin(2\pi f_1 \cdot t_{1\max})}{2\pi f_1 \cdot t_{1\max}} \frac{\sin(2\pi f_2 \cdot t_{2\max})}{2\pi f_2 \cdot t_{2\max}} ... \frac{\sin(2\pi f_N \cdot t_{N\max})}{2\pi f_N \cdot t_{N\max}}. \quad (26)$$

Thus, finite acquisition time causes a convolution of NMR spectrum with *sinc* function. This manifests itself in peak broadening and presence of *sinc* "wiggles". The broadness of the NMR peak is thus dependent not only on relaxation rate but also on the maximum evolution time. Both effects correspond to the Fourier Uncertainty Principle [53] stating that, in general, the "broadness" of time representation and frequency representation are inversely proportional to each other.

3.2.3 Sampling

An NMR signal is measured in a discrete manner, i.e., sampled. This may be represented by a *sampling function*, being a multidimensional train of K delta pulses:

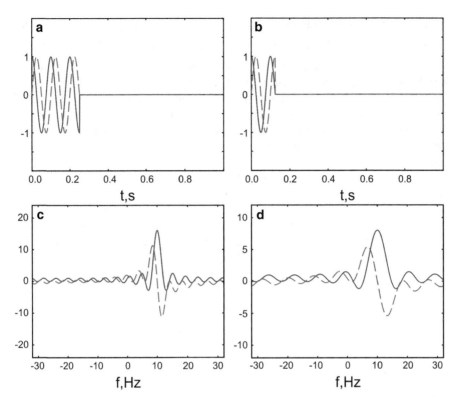

Fig. 4 Signal truncation and spectral line width (real and imaginary parts marked with *solid* and *dashed lines*, respectively). (**a**) Signal truncated to 250 ms. (**b**) Signal truncated to 125 ms. (**c**) Spectrum of a signal truncated to 250 ms – *sinc* function. (**d**) Spectrum of a signal truncated to 125 ms – *sinc* function

$$III(t_1, t_2 ... t_m) = \sum_{k=1}^{K} \delta\left(t_1 - t_1^k, t_2 - t_2^k, ..., t_N - t_N^k\right). \quad (27)$$

Discrete sampling can be thus represented by multiplication of continuous signal by $III(t_1, t_2 ... t_m)$.

In general, one can distinguish between two kinds of sampling: *uniform* (or conventional), i.e., with sampling coordinates $\left(t_1^k, t_2^k, ..., t_N^k\right)$ corresponding to full Cartesian grid and *non-uniform*, i.e., with coordinates chosen arbitrarily, according to one of the sampling schedules (see Sect. 4).

Conventional discrete sampling influences a spectrum in quite a straightforward way because of the simplicity of the corresponding Fourier pair:

$$III\left(\frac{t_1}{\tau_1}, \frac{t_2}{\tau_2} ..., \frac{t_m}{\tau_m}\right) \underset{\text{IFT}}{\overset{\text{FT}}{\longrightarrow}} III(\tau_1 f_1, \tau_2 f_2 ..., \tau_m f_m). \quad (28)$$

As a result, an infinite number of "copies" are produced in spectral domain, with the distance $\left(\frac{1}{\tau_1}, \frac{1}{\tau_2}, \ldots, \frac{1}{\tau_m}\right)$ between "copies" (see Figs. 5 and 6a, b). If the distance is greater than half the signal bandwidth, then copies do not overlap. This leads to the well known Shannon–Nyquist Sampling Theorem [54], which says that the spectrum of a band-limited signal can be perfectly recovered from discrete samples if sampling frequency is at least two times higher than highest frequency present in the signal. Obviously, if the spectrum is perfectly recovered, then a *continuous* signal is recovered as well, meaning that discrete points are interpolated with *sinc* functions.

Usually, the sampling frequency is set based on predicted spectral width of a signal. Assuming that the frequency band is limited, one can take the central, low-frequency spectrum "copy" that is equivalent to the perfect spectrum (see Fig. 5b). Actually, this is done by default by standard FT algorithms, e.g., Fast Fourier Transform *(FFT)*. In the case, when the prediction is wrong, i.e., sampling interval is higher than required by the Shannon–Nyquist criterion, the "copies" of spectrum overlap, which leads to the phenomenon known as *peak folding* or *aliasing*. This

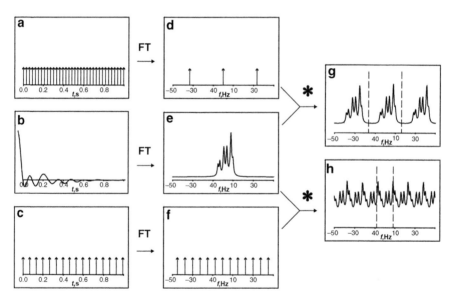

Fig. 5 Aliasing phenomenon. (a) Train of delta pulses representing sampling with $\tau = 0.03$ s (sampling rate 33.33 Hz). (b) Continuous signal – multi-component oscillatory function. (c) Train of delta pulses representing sampling with $\tau = 0.066$ s (sampling rate 15 Hz). (d) Train of delta pulses in frequency domain being FT of sampling schedule (a). (e) Spectrum of continuous signal (b). (f) Train of delta pulses in frequency domain being FT of sampling schedule (c). (g) Convolution of (d) and (e) corresponding to FT of signal (b) sampled according to (a). Properly sampled spectral bandwidth with central spectrum "copy" between *dashed lines*. No aliasing. (g) Convolution of (f) and (e) corresponding to FT of signal (b) sampled according to (c). Properly sampled spectral bandwidth with central spectrum "copy" between *dashed lines*. Aliasing due to insufficient sampling rate

phenomenon manifests itself by the presence of peaks at false frequency coordinates in the spectral region of interest.

It should be noted that an upper limit for the sampling interval results in a lower limit for peak width. This becomes especially significant in the multidimensional NMR experiments, where each sampling point takes a few seconds of experimental time (see Scheme 1). Moreover, the requirements of regular sampling grow exponentially with the number of dimensions and, despite measurements over hours or days, natural, relaxation-determined peak widths are rarely obtained even in 3D spectra.

Coupling between peak width and number of sampling points (i.e., experimental time) is the main reason for the use of non-uniform sampling in NMR.

For non-uniform sampling the FT of sampling schedule (sometimes referred to as *Point Spread Function, PSF*) is not a simple train of delta pulses, but becomes a more complex function (see Sect. 4 and Fig. 6). This, in general, leads to three conclusions:

1. For arbitrary, non-uniform sampling it is no longer possible to obtain a spectrum that is equal to a spectrum of continuous signal, even if it is strictly band-limited. Spectral artifacts, depending on the sampling schedule, appear as part of Point Spread Function.
2. Aliasing, however, occurs in the presence of a sampling grid, from which sampling points are taken. For purely off-grid sampling aliasing does not appear (notably, NMR hardware allows very fine approximation of off-grid sampling). This means that one can use non-uniform sampling to remove coupling between sampling rate and line width and obtain high spectral resolution in a relatively short experimental time (see Sect. 8).
3. The negative effect of non-regular sampling, i.e., presence of artifacts, is separate from all other spectral effects associated with various signal features,

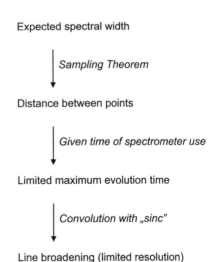

Scheme 1 Scheme illustrating limitations associated with conventional sampling, i.e., coupling between experimental time and line width

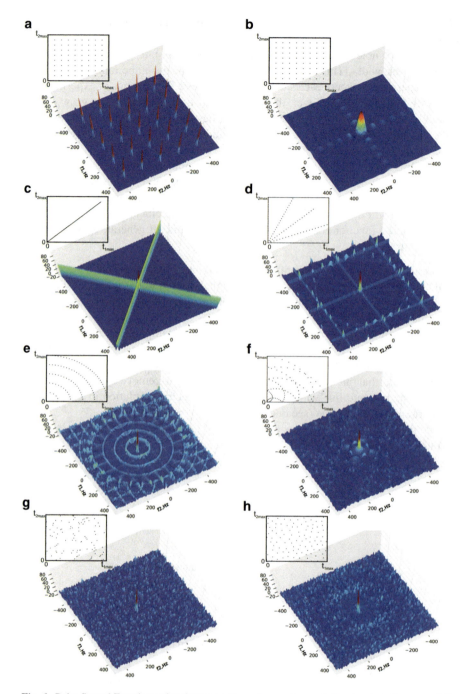

Fig. 6 Point Spread Functions of various sampling schemes (presented on insets in the *upper left* part of each panel): (**a**) conventional sampling ($t_{1\mathrm{max}} = t_{2\mathrm{max}} = 50$ ms), (**b**) conventional sampling ($t_{1\mathrm{max}} = t_{2\mathrm{max}} = 5$ ms), (**c**) radial sampling with one sampling line, (**d**) radial sampling with

e.g., relaxation, amplitude, etc., and can be discussed independently. In other words, FT of an irregularly sampled signal has the same features as FT of a conventionally sampled signal, differing only in PSF.

4 Non-Uniform Sampling Schemes

Conventional (Cartesian grid) sampling as a scheme is an obvious method of choice, when experimental time and/or line width expense is acceptable. However, when especially narrow spectral lines or high dimensionality are required, irregular sampling should be employed. As stated above, it can make peak widths independent of experimental time. Nevertheless, one should always remember about cost of irregularity, i.e., introduction of spectral artifacts, whose pattern and level is dependent on a sampling scheme.

4.1 Radial Sampling

Radial sampling was the first sparse sampling scheme introduced into NMR. Apart from FT [44, 55, 56], other data processing techniques were proposed. These include reduced dimensionality [57], projection-reconstruction [41], and multi-way decomposition [58]. Radial sampling scheme consists of points placed on a set of lines in the time domain (see Fig. 6c, d). The coordinates of the ith point, lying on the jth line, can be described in a polar coordinate system as

$$t_1 = i \cdot \Delta r \cdot \cos \psi^j, \tag{29a}$$

$$t_2 = i \cdot \Delta r \cdot \sin \psi^j. \tag{29b}$$

PSF of radial distribution is a set of ridges (see Fig. 6c, d). Each pair of ridges is a FT of one sampling line and they are oriented at ψ^j and $\psi^j + \pi/2$ angles.

Fig. 6 (continued) five sampling lines, (**e**) concentric rings sampling, (**f**) spiral sampling, (**g**) purely random sampling, (**h**) Poisson disk sampling. Number of points in each sampling scheme is equal to 100. For panels (**c**)–(**h**) $t_{1max} = t_{2max} = 50$ ms

4.2 Concentric Rings Sampling

Concentric rings sampling was proposed by Coggins and Zhou [59]. The sampling scheme is depicted in Fig. 6e. Importantly, the number of points situated on each ring increases linearly with a ring's radius (linearly increasing concentric ring sampling, LCRS). The coordinates of the ith time point, lying on jth ring are

$$t_1 = r \cdot \cos(i \cdot \Delta \psi^j + \psi_0^j), \tag{30a}$$

$$t_2 = r \cdot \sin(i \cdot \Delta \psi^j + \psi_0^j). \tag{30b}$$

In LCRS ψ_0^j is the same for all rings. If this phase is chosen randomly for each ring independently, the scheme is called RLCRS (randomized LCRS). PSF of LCRS takes the form of a set of ring-shaped ridges, and for RLCRS this pattern is slightly disturbed, covering the spectral space more evenly.

4.3 Spiral Sampling

Point coordinates in the spiral sampling scheme [44] are defined as

$$t_1 = i \cdot \Delta r \cdot \cos(i \cdot \Delta \psi^j), \tag{31a}$$

$$t_2 = i \cdot \Delta r \cdot \sin(i \cdot \Delta \psi^j). \tag{31b}$$

Notably, both radial and LCRS schemes can be considered as special cases of the spiral sampling scheme. PSF of this distribution is the combination of the two above PSFs. The artifacts form ring-shaped ridges, but the intensity is not constant along the rings, but varies with angle (see Fig. 6f) [60].

4.4 Random Sampling

As can be clearly seen from the above examples, regularity in time domain results in regularity in frequency domain. This suggests that very irregular, random sampling schemes can be particularly useful (see Fig. 6g). In this case the artifacts are "spread" evenly over the spectral space; their level is thus reduced, comparing to more regular sampling schemes, and consequently it is less probable that false peaks will come up.

Optimization of such purely random distribution can be done by introducing certain constraints which protect against choosing one point too close to another.

A few algorithms of generation of such semi-random sampling schemes were investigated [60, 61]. Among them, Poisson disk sampling was found to be the most optimal (see Fig. 6h). It directly assumes a minimal distance between time points. The artifact level is not as even as in a purely random case. It is lower in the vicinity of the peaks. Moreover, by slight modification of the restraints, one can adjust the shape of the "clean" region to spectral widths or to compensate for different maximum evolution times in different dimensions. Another variant of distance-restrained sampling, referred to as Poisson-gap sampling, was presented by Hyberts and coworkers and used with forward maximum entropy processing [62].

4.5 Weighted Samples and Weighted Probability

In a conventional case, usually a certain weighting function is applied in each dimension for improving signal-to-noise ratio or reducing the effect of signal truncation (*sinc* "wiggles"). This procedure is called Weighted Samples (WS) and can also be performed for random sampling schemes. In the case of irregular sampling, however, an alternative solution can also be employed. Instead of applying weighting function to a sampled signal, the function may be used as a probability distribution during generation of a randomized sampling scheme. Such an approach is referred to as Weighted Probability (WP). The two procedures (WS and WP) result with spectra of the same line shape and S/N for a non-decaying signal (i.e., if S/N is constant in time) [63]; see Fig. 7. However in the case of a real, relaxing, and noisy FID signal, the WP method is more effective, as more points of higher S/N (from the beginning of FID) are measured [60].

5 Methods of Integration

FT is an integral operation. In a real case the transformed function is discrete, and thus the integral has to be replaced by a sum. Therefore, for irregular points distributions, unequal distances between sampling points can be taken into account by applying certain weights. In analogy to 1D numerical integration employing a rectangular or trapezoidal rule, for the multidimensional case one can obtain weights by Voronoi tessellation [60, 64] or Delaunay triangulation [46]. These methods, although helpful in the case of polynomials integration or rational functions, are rather unsuitable for rapidly oscillating functions. When the sampling density is lower than density determined by Sampling Theorem (i.e., always, when non-uniform sampling is justified), introduction of mentioned weights diminishes the signal-to-artifact ratio (see Fig. 8). Therefore, for oscillatory functions, simple summation is more appropriate, i.e., Monte Carlo integration [65, 66]. It does not affect signal-to-noise ratio and the result converges to the exact value with $\sqrt{n}$.

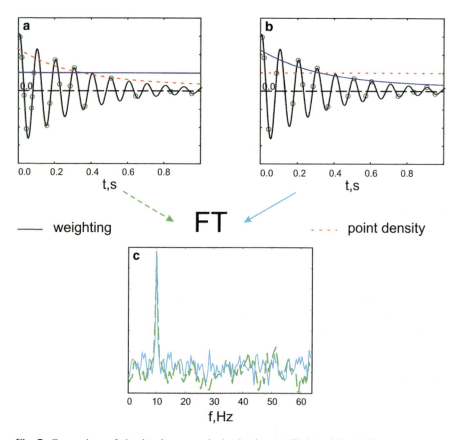

Fig. 7 Comparison of simulated spectra obtained using (**a**) Weighted Probability (exponential distribution, *dashed line* on panel **c**) and (**b**) Weighted Samples (exponential weighting, *solid line* on panel **c**) methods. The signal was simulated without thermal noise. Both methods give spectra with the same signal to artifact ratio and line widths

6 Sparse Sampling and FT as a Linear Algebra Issue

One-dimensional FT of a sampled signal (or its spectrum) may be thought of as a solution to a system of equations:

$$\frac{1}{\sqrt{N}} \begin{bmatrix} e^{i2\pi f^1 t^1} & e^{i2\pi f^2 t^1} & \ldots & e^{i2\pi f^m t^1} \\ e^{i2\pi f^1 t^2} & e^{i2\pi f^2 t^2} & \ldots & e^{i2\pi f^m t^2} \\ \ldots & \ldots & \ldots & \ldots \\ e^{i2\pi f^1 t^n} & e^{i2\pi f^2 t^n} & \ldots & e^{i2\pi f^m t^n} \end{bmatrix} \begin{bmatrix} S(f^1) \\ S(f^2) \\ \ldots \\ S(f^m) \end{bmatrix} = \begin{bmatrix} s(t^1) \\ s(t^2) \\ \ldots \\ s(t^n) \end{bmatrix}. \tag{32}$$

Or, more briefly:

$$\hat{A}\vec{S} = \vec{s} \tag{33}$$

Generalized Fourier Transform for Non-Uniform Sampled Data

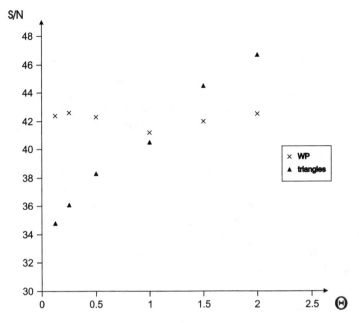

Fig. 8 The plot of spectral signal-to-artifact ratio of simulated spectrum $f(t_1, t_2) = \exp(-2\pi i_1 \cdot 300 \text{ Hz} \cdot t_1 - 50 \text{ Hz} \cdot t_1 - 2\pi i_2 \cdot 300 \text{ Hz} \cdot t_2 - 50 \text{ Hz} \cdot t_2)$ in function of relative density of time domain points ($\Theta = \rho/\rho_N$) comparing WP method and surface integration procedure (512 evolution time points of Gaussian PDF: $\exp(-t^2/2\sigma^2), \sigma = 0.5$). Spectral widths and maximum evolution times were equal: $sw_1 = sw_2, t_{1\max} = t_{2\max} = t_{\max} = 0.02$ s. Θ was changed by varying both spectral widths (and consequently ρ_N) keeping constant number of points and evolution time surface $t_{\max}^2$ (and ρ consequently). Reprinted with permission from [46]

where $\hat{A}$ is an inverse FT matrix with number of rows n equal to the number of time points and number of columns m equal to the number of frequency points, $\vec{S}$ is an m-element vector representing spectrum, and $\vec{s}$ represents vector of n signal samples:

$$A_{ij} = \frac{e^{i2\pi f^j t^i}}{\sqrt{N}}, \tag{34a}$$

$$S_i = S(f^i), \tag{34b}$$

$$s_j = s(t^j). \tag{34c}$$

Thus, the usual spectral processing task is to find unknown $\vec{S}$ that agrees with known $\vec{s}$ (fulfils the system of equations). The possible situations are:

1. $n = m$ and matrix $\hat{A}$ is full rank. Then the system of equations has a unique solution and it may be obtained by multiplying both sides by FT matrix, which is a Hermitian transpose (conjugate transpose) of matrix $\hat{A}$:

$$\vec{S} = \left(\hat{A}^*\right)^T \vec{s}. \qquad (35)$$

In the case of equally spaced sampling, matrix $\hat{A}$ is highly symmetric and FFT algorithms may be employed to reduce computational time.

The number of samples taken from the signal (n) determines the number of frequency points that are possible to be determined. Adding zero-valued, "artificial" sampling points at the end of the signal allows one to calculate the increased number of spectral points. This procedure (known as *zero filling*) is the equivalent of interpolation in the spectral domain [67].

2. $n > m$. The system of equations is overdetermined and strictly speaking there is no solution. However, the number of equations may be reduced and the solution can be obtained with additional gain on signal-to-noise ratio. This is achieved by various digital filtering techniques. The situation corresponds to *oversampling* and in practice exists only in directly detected signal.
3. $n < m$. The system of equations is underdetermined and there are many possible solutions. This corresponds to sparse, non-uniform sampling.

Among the spectral vectors, that fulfil the system of equations, there is an optimal one, i.e., spectrum of a signal sampled in a uniform manner. Finding it, however, is not a simple task (even if thermal noise could be neglected). Many approaches were presented, differing in type of constraints that limits the number of solutions. Some of these include the following:

Maximum Entropy Methods – the solution with highest entropy is found. Various "entropy" functions were used in the past [49, 68].

Integration of frequency and time domain information [69] by assuming that some of the frequency points are equal to zero.

l_1-*norm minimization*, the solution with smallest l_1-norm (sum of absolute values of spectral points) is found. It was proved recently that for signals featuring "dark" spectra (small number of non-zero frequencies) l_1-norm should lead to an optimal solution by convex minimization [70]. This approach has been successfully employed in many branches, including MRI [71, 72]. So far, only a few examples of use of this method in NMR spectroscopy were presented [73, 74], but l_1-norm penalty function has been used before in various NMR processing tasks [75, 76].

Interpolation or gridding of sparse dataset may help to recover missing data points and use conventional FT processing. This, however, may lead to significant disturbances if the simplest, polynomial interpolation is used [77]. More advanced gridding techniques are helpful here [78].

Non-uniform FT (nuFT) employing an equation identical to (35), but with non-quadratic matrix $\hat{A}$ (or with full $m \times m$ matrix $\hat{A}$ and zeros at non-sampled points of m-long vector $\vec{s}$). The obtained solution features minimum l_2-norm (power), which can be easily proved, considering that FT is an unitary operation and thus l_2-norms of signal and spectrum are equal (Parseval's theorem). Although, the solution is not the optimal one, the processing is fast and was successfully employed in many applications [79–84]. Moreover, the spectrum can be additionally improved by application of various artifact-cleaning algorithms [60, 83, 85].

7 Suppression of Sampling Artifacts in FT Spectra

7.1 The Principle of CLEAN Algorithm

7.1.1 The Model of "Dark" Spectrum

As mentioned in Sect. 6, in the case of sparse sampling there are insufficient data to determine uniquely the Fourier representation of the measured signal. Therefore, sampling artifacts observed in nuFT spectra can be regarded as an unavoidable consequence of missing data. However, more accurate spectral estimates can be obtained by incorporating a priori knowledge about the nature of sampled signal. For a certain class of signals it might be assumed that the continuous Fourier spectrum consists of a small number of well-localized components (peaks) and relatively weak flat (frequency independent) noise:

$$\hat{S}(\vec{f}) = \sum_i \hat{S}_i(\vec{f}) + \varepsilon(\vec{f}). \qquad (36)$$

This general model, usually referred to as "dark" spectrum, allows a variety of reconstruction methods to be employed (see Sect. 6). The CLEAN algorithm, proposed originally for the reconstruction of two-dimensional maps in radio astronomy [86], utilizes essentially the same signal properties. It is noteworthy that the "dark spectrum" model is especially well suited to multidimensional NMR spectroscopy.

7.1.2 Description of the CLEAN Procedure

The starting point of the procedure is a discrete FT spectrum:

$$\hat{S}(\vec{f})^{(0)} = \hat{S}(\vec{f}) = \text{FT}[s(\vec{t}) \cdot III(\vec{t})], \qquad (37)$$

which is a convolution of continuous spectrum with the FT of sampling function (Point Spread Function). The latter is usually termed "dirty mask" in this context.

The aim of the procedure is to identify well-localized sources of artifacts present in FT spectra. Intuitively, one may suppose that they can be found by computing the convolution of FT spectrum and the PSF:

$$\hat{C}(\vec{f}) = \hat{S}(\vec{f}) * \tilde{III}(\vec{f}). \qquad (38)$$

Following the convolution theorem (see Sect. 2.3) one shows that this convolution is the FT spectrum itself:

$$\hat{C}(\vec{f}) = \hat{S}(\vec{f}) * \tilde{III}(\vec{f}) = FT[\hat{s}(\vec{t}) \cdot III(\vec{t}) \cdot III(\vec{t})] = FT[\hat{s}(\vec{t}) \cdot III(\vec{t})] = \hat{S}(\vec{f}). \quad (39)$$

Not surprisingly, it appears that one can use discrete FT spectrum to find the most probable sources of spectral artifacts.

In the first step, one shifts the centre of PSF (normalized to one at maximum) to the point where $\hat{S}(\vec{f})$ has a maximum absolute value $|I_{max}|$ (see also Fig. 9). Then one subtracts a fraction $0 < \gamma \leq 1$ (called "loop gain") of the shifted PSF:

$$\hat{S}(\vec{f})^{(i+1)} = \hat{S}(\vec{f})^{(i)} - \gamma \tilde{III}(\vec{f} - \vec{f}_{max}). \quad (40)$$

The extracted component gives rise to the so-called "replica", $G(f)$, which is hoped to reproduce the perfect spectrum $S(f)$ at the end of the procedure:

$$\hat{G}(\vec{f}_{max})^{(i+1)} = \hat{G}(\vec{f}_{max})^{(i)} + \gamma \cdot I_{max}. \quad (41)$$

Providing that the peak at the selected point was a real feature, one obtains a new spectrum $\hat{S}(\vec{f})^{(i+1)}$ with decreased level of artifacts.

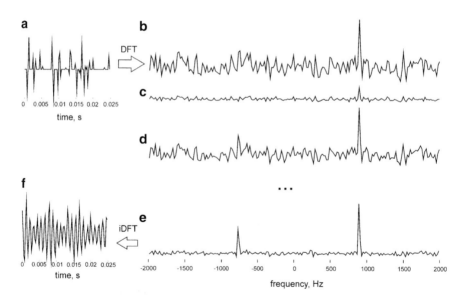

Fig. 9 The principle of the CLEAN algorithm visualized on a simulation of three signals of relative amplitudes 1:5:10 and equal decay rates. The sparsely sampled signal (**a**) is Fourier transformed (**b**), then the mask (**c**) is subtracted to yield the residual spectrum. Reconstruction after the first iteration is shown (**d**). The final result of the CLEAN procedure (**e**) can be used to obtain reconstruction of the time-domain signal (**f**)

In the next iteration, one can repeat the steps of (1) finding the most intense spectral amplitude and (2) subtraction of shifted PSF. The whole procedure should be continued until there are no significant peaks in the spectrum. This condition can be formulated as follows:

$$|I_{\max}| < \alpha \cdot \sigma_i, \qquad (42)$$

where σ_i is the estimated noise level in the ith iteration, and α is usually a small integral value (3–5). It should be noted that σ_i is a measure of both remaining artifacts and usual thermal noise $\varepsilon(\vec{f})$.

Finally, the residual spectrum may be added to "replica" in order to retain smaller features that might have been omitted by the CLEAN algorithms, or to reintroduce the usual noise $\varepsilon(\vec{f})$. The latter might be useful to judge which peaks selected during the iterations are false [87]. It was emphasized that displaying the "replica" without the addition of residual spectrum is merely a "cosmetic" operation and does not improve the sensitivity at all [88].

It is noteworthy that the uncertainty of peak amplitudes caused by the presence of noise $\varepsilon(\vec{f})$ limits the capability of CLEAN algorithm to improve the quality of spectrum [89]. This, however, should apply for the most of reconstruction algorithms, e.g., similar conclusions were drawn for the maximum entropy method [90].

7.1.3 Discussion of the Parameters of CLEAN

Apparently, CLEAN has two parameters which can affect both efficiency (in terms of computational effort) and accuracy of the final results. *Loop gain*, γ, determines how fast the artifacts are removed from the spectra in each iteration. Generally, it should reflect the probability that the selected peak is true (neither an artifact nor a noise peak) and that the intensity observed in the spectrum $I_{\max}$ comes entirely from the component centered at $\vec{f}_{\max}$. Therefore, small values should be used for spectra containing overlapped peaks [91], peaks broader than PSF [87, 91], or noisy ones. Alternatively, *loop gain* can take a variable value depending on the ratio $|I_{\max}|/\sigma_i$. Clearly, only the infinitesimally small value of loop gain ensures maximum safety of the procedure [86]. However, decreasing the loop gain causes a serious efficiency penalty, and the compromised values between 0.25 and 0.5 are typically used [88]. It has been pointed out that larger values can result in false splittings, especially when there is a mismatch of the "mask" line widths and the experimental ones [91].

The second parameter is the intensity threshold, which can be determined by the operator in advance and kept fixed, or evaluated dynamically on the basis of the current noise level in the spectrum. The former option requires prior knowledge of noise amplitude, whereas the latter needs a robust method of measuring the noise level.

Regarding the termination criterion, one should comment that there is a trade-off between the safety of peak identification and completeness of artifact suppression [91]. For example, the threshold of $5\sigma_i$ provides great confidence that only genuine peaks are extracted; however, it also limits the benefits of the CLEAN procedure as the artifacts originating from less intense components remain in the spectrum.

It has also been agreed that a fixed number of iterations is difficult to apply in practice for NMR spectra and could lead to misinterpretations of the results of CLEAN algorithm [89]. Therefore, one should rather use the intensity threshold as the stopping condition.

Other authors [88] also noted that it is advantageous to use fine digitization in the frequency domain as it enables one to position precisely the "mask" (PSF). On the other hand, this does not seem critical for the results and may unnecessarily increase the computational burden.

7.2 Development of the CLEAN Algorithm

As noticed by Coggins and Zhou [83], if CLEAN is employed to suppress artifacts originating from irregular sampling, the artifact level varies greatly in the multidimensional spectrum along a directly detected dimension. Consequently, it is impractical to use a fixed intensity threshold in this case. Apart from the commonly used dynamic threshold of $5\sigma_i$, it was suggested to employ the noise stabilization criterion, which stops the iteration if CLEAN no longer efficiently removes artifacts. The condition was quantified as follows:

$$\bar{\sigma}_j \leq (1+\tau)\bar{\sigma}_i \quad \text{for} \quad i-25 \leq j < i, \tag{43}$$

where $\bar{\sigma}_j$ denotes the average noise level measured in the jth iteration. The tolerance for noise stabilization τ of approximately 0.05 was suggested. One may consider this condition a practical optimization of CLEAN algorithm in terms of numerical efficiency, not necessarily improving the quality of the final spectra.

In contrast to other implementations, Coggins and Zhou used the mask computed directly from the sampling function, without the knowledge of minimal signal line width. It was argued that such approach is more general, as broad or overlapped peaks can be represented sufficiently accurately by a superposition of narrow peaks. Indeed, frequently the resolution of spectra of biomolecules is mostly determined by signal truncation, and the natural peak line widths can be neglected when using CLEAN in these applications.

A different approach to CLEAN processing was suggested by Kazimierczuk and co-workers [60]. In their implementation, peaks are manually fitted in the initial spectrum using assumed shapes (e.g., Lorentzian or Gaussian, depending on the decay of sampling density employed). The list of peaks and their line widths, which can be considered the analytic equivalent of "replica", are then provided to the processing program. In the following, the artifacts generated by the peaks in the list are computed and subtracted from the initial spectrum, except for the peak positions and their vicinities. The procedure can be repeated if advantageous, e.g., if a significant number of (new) medium and small peaks were found after subtraction of the artifacts.

It should be noted that the use of an appropriate analytic function instead of discrete "replica" may be beneficial as this (1) is less influenced by noise and (2) allows one to reproduce the "wings" of the resonances, which are neglected in the original CLEAN algorithm due to intensity threshold. On the other hand, if the line widths in the Fourier domain are mainly due to signal truncation, the fitted parameters poorly reflect the true signal properties, and this may affect the performance of described procedure.

A remedy for this was proposed by Stanek and Koźmiński [85]. In their significantly modified version, referred to as Signal Separation Algorithm (SSA), peaks are automatically found and fitted using the mono-exponentially decaying functions in the time domain. As a consequence, the simulated line shapes in the Fourier domain are affected by the sampling process in the same manner as the real peaks in the spectrum. The advantages of this approach over the original CLEAN were demonstrated [85] (see also Fig. 10). Another modification proposed by these authors regards the case of overlapped peaks or when decay parameters cannot be reliably established. It was suggested to find a replica that reproduces the observed peak shape in the iterative process. The idea to vary the amplitudes in replica until the desired peak shape is obtained clearly alleviates the problem of the appropriate value of *loop gain*.

7.3 The Early Applications of CLEAN to NMR Spectroscopy

The original CLEAN algorithm was invented to deconvolve effectively the Fourier spectrum from the PSF. In radio astronomy it was either impossible or impractical to arrange detectors on a regularly spaced grid due to malfunctioning of the part of equipment, occultations caused by the Moon, or if telescopes were operating over a

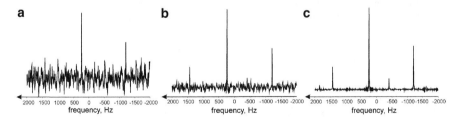

Fig. 10 Comparison of the efficiency of CLEAN (**b**) and SSA (**c**) shown on a simulated signal containing six components of relative amplitudes 1:2:4:8:16:32 and equal decay rates of 20 s^{-1}. Additionally, white Gaussian noise of $\sigma = 0.02$ was present. Both algorithms started from the same initial nuFT spectrum (**a**), and the same threshold for peak detection equal to $5\sigma_i$ was used. Spectral width of 4 kHz and max. evolution time of 70 ms were set. Seventy out of 280 points were sampled, yielding a relative sampling density of 0.25

large area to provide high resolution maps [86]. The aim of CLEAN was to convert the map obtained from an irregular and/or coarse grid of interferometers to that which would be obtained from a fine and complete grid.

As pointed out by Davies and co-workers [91], many high-resolution 2D spectra are not sparse and suffer rather from signal overlap or line shape distortions occurring for several reasons (twisted-shape, truncation artifacts, inhomogeneous broadening). Although these difficulties seem quite different from those in radio astronomy, it became possible to employ essentially the same algorithm to alleviate these problems.

The idea was to construct the "mask" similar to an undesired shape observed in the spectra, and use CLEAN to replace distorted peaks with those of a perfect Lorentzian shape. Shaka and co-workers [88] showed that the algorithm is able to convert a twisted shape to a double-absorption in 2D phase-sensitive J spectra of complex organic molecules. This was achieved by (1) locating the twisted-shape peaks, (2) simulating double-dispersion signals of the same line width at the same frequency coordinates, and (3) subtraction of the latter from the original spectrum. Effectively, the most intense peaks were in double-absorption while those ignored by CLEAN remained in the twisted shape.

A similar approach was presented by Keeler in application to heteronuclear J spectra with highly truncated echo modulation [87]. Truncation of signal, used for sensitivity reasons, results in "sinc wiggles". These artifacts can be suppressed by apodization, although at the expense of resolution. Keeler showed that CLEAN is an inexpensive alternative to the maximum entropy method, which can also remove truncation artifacts without degrading resolution.

The difficulty that has arisen in both applications was to adjust the line width of the mask in order to fit all signals. It has been suggested that, if there is a mismatch of the line widths between the mask and experimental line shapes, one has to decrease *loop gain* and represent broad peaks as a superposition.

Davies and co-workers alleviated the problem of the optimal mask, by using experimental line shape of a well separated singlet resonance [91]. This was showed to enhance resolution of spectra more effectively than when a simulated Lorentzian mask is employed. Additionally, CLEAN was compared to maximum entropy method, giving similar results in a considerably shorter computational time. One should note that the use of experimental aperture shape to compensate for spatial inhomogeneity of a magnetic field is limited to the cases of high S/N, otherwise the mask is heavily biased by noise.

In all the cases described above the fixed threshold (of a few per cent of the tallest peak) was used to terminate processing. This was possible as the noise level does not significantly vary during the iterations. The latter does not hold in the case of irregular sampling and more careful termination criteria have to be applied when deconvolving the PSF [89]. It has been showed on both experimental data and simulations that similar results can be obtained by CLEAN and maximum entropy method, and that CLEAN performs much better in recovery of missing samples than in extrapolation of a truncated signal [89]. As mentioned above, the success of CLEAN was limited by S/N.

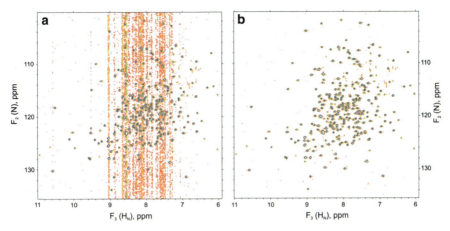

Fig. 11 F_2/F_3 projections (along $F_1(C')$ dimension) of 3D HNCO-TROSY spectra for maltose binding protein (371a.a., uniformly-deuterated, 0.5 mM D_2O/H_2O 1:19 solution), obtained using sparse on-grid sampling, nuFT (**a**), and SSA processing (**b**). The data were recorded at the Varian 700 MHz spectrometer, assuming the spectral widths of 2.8 and 2.5 kHz in $F_1(C')$ and $F_2(N)$ dimensions, respectively. A total of 1,750 sampling points were generated using decaying sampling density ($\exp(-t^2/2\sigma^2)$, $\sigma = 0.5$). Maximum evolution times of 30 and 50 ms were set, yielding a relative density of $\theta = 16.7\%$

According to these observations, the power of CLEAN algorithm was utilized in high-dimensional (3D and 4D) NMR spectroscopy of proteins (see Fig. 11), where *sparse sampling* has to be employed due to practical limitation on experiment time (see Sect. 9.4).

In conclusion, the application of CLEAN algorithm to sparsely sampled data is especially beneficial if (1) the technique features good thermal sensitivity and (2) a high dynamic range of peak amplitudes is expected. Otherwise, artifact suppression is hampered or irrelevant in view of the general noise level.

7.4 Algorithms Related to CLEAN

It is noteworthy that the principle of CLEAN algorithm was also utilized in several other processing methods [40, 75]. Kupče and Freeman adapted the processing scheme to remove ridges and false peaks present in the projection-reconstruction of 3D spectra [40]. As noted by the authors, it can be confidently assumed that the tallest peak in the reconstruction is genuine. It is then possible to extract it *from the projections* and reconstruct the full spectrum again. As usual, the process can be repeated to suppress projection-reconstruction artifacts further until no significant peaks are present. At the final stage, the extracted peaks are reintroduced to the full spectrum.

Hyberts and co-workers described a "distillation" procedure which improves the quality of Forward Maximum entropy (FM) and l_1-norm reconstructions [75]. The purpose of this processing scheme is to divide FID signal into two components, one containing "tall" and another "small" spectral information. The division is performed in the Fourier domain according to the relative amplitude of each pixel to the most intense one, and both parts "small" and "tall" are inversely transformed to the time domain. The advantage is that FM performs the reconstruction on a sub-spectra of decreased dynamic range of peak amplitudes. This was shown to improve both linearity of the method and suppression of sampling artifacts. This is in analogy to CLEAN processing, where time domain signal is effectively split to the contributions from strong and weak signals. It is noteworthy that the "distillation" procedure does not require any parameters and usually up to eight iterations are sufficient.

8 FT as a Tool for Large Evolution Time Domain

8.1 Features of "Sampling Noise"

As mentioned above, the nuFT does not find the optimal solution that fits to the experimental data. Spectra obtained by nuFT suffer from additional artifacts, which, in the case of random sampling, take a noise-like form. Luckily, they also reveal similar properties like thermal noise, i.e., the artifact level is proportional to $\sqrt{N}$ (see Fig. 12) and does not depend on a dimensionality of a signal, maximum evolution times nor spectral widths [45]. This fact may be proved in various ways (two of them were presented in [45]), and below we will present a new, simpler proof, based on known properties of Monte Carlo integration [66].

Discrete multidimensional FT of randomly sampled signal may be considered as an estimation of continuous multidimensional integral (12) with Monte Carlo procedure. According to properties of Monte Carlo integration, associated with

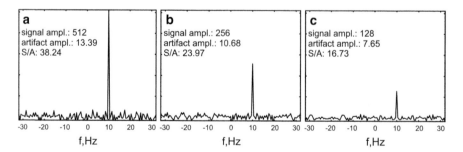

Fig. 12 Peak amplitude, artifact level, and signal-to-artifact ratio for spectrum of non-decaying signal of frequency 10 Hz, sampled with: (**a**) 512, (**b**) 256, and (**c**) 128 points. Uniform random sampling was used

the law of large numbers and the central limit theorem, the approximate value of an integral with some finite integration volume V:

$$I = \int_V f(\bar{x})d\bar{x} \qquad (44)$$

is given by

$$Q = \frac{V}{N}\sum_{i=1}^{N} f(\bar{x}_i), \qquad (45)$$

with an expected value equal to the value of a continuous integral (unbiased estimator):

$$E[Q] = I, \qquad (46)$$

and the variance decreasing with N:

$$\text{var}(Q) = V^2 \frac{\text{var}(f(\bar{x}))}{N}. \qquad (47)$$

Thus, the error of approximation is decreasing with $\sqrt{N}$. The only difference between signal processing and numerical integration is the way samples are obtained. Instead of calculating the values of function at randomly selected points, as is done in the Monte Carlo procedure, the integrated function is *experimentally measured* at these points (or, more strictly, measured and multiplied by the transform kernel). Nevertheless, the way points are obtained does not affect general conclusions, i.e., that the estimator is unbiased and converges to the perfect, artifact-free spectrum with a growing number of sampling points and that the relative error of the result (S/A ratio) is inversely proportional to $\sqrt{N}$. Notably, the estimation error does not depend on parameters that cause sampling-related problems (i.e., limited resolution) in a conventional approach, e.g., dimensionality of a signal and maximum evolution time (see Fig. 13). This feature makes random sampling a perfect tool for high-dimensional (4D, 5D, 6D, etc.) NMR experiments [45] with quite high absolute numbers of sampling points (not necessarily meaning high sampling density!). For the same reasons, Monte Carlo is known to be a favorable method for integration of high-dimensional functions [66]. It is also noteworthy that other features of random sampling processed with nuFT have their equivalents in the Monte Carlo method. For instance, stratified sampling is known to reduce variance of an integral estimation [66].

Besides the absolute number of points, the artifact level is also inherently associated with a number of peaks and their intensities (as artifacts are "part" of the Point Spread Function, see Fig. 14). Thus, the more peaks in a spectrum, the

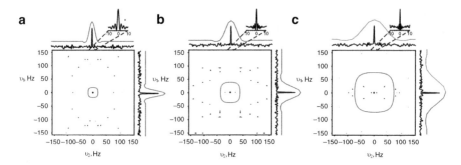

Fig. 13 2D cross-sections from simulated spectra: (**a**) 3D, (**b**) 4D, (**c**) 5D. The threshold was set at 10% of peak intensity; 256 time points were generated randomly with uniform distribution and maximum evolution time of 0.4 s (**a**), 0.8 s (**b**), and 1.6 s (**c**), in all dimensions. The distance between spectral points was set to the reciprocal of maximum evolution time in order to hide the effect of signal truncation. The *insets* show a spectral line narrowing obtained by MFT using higher digital resolution. Simulation was repeated for the conventional set of 256 points, with the Nyquist rate of 16 × 16 (**a**), 8 × 8 × 4 (**b**), and 4 × 4 × 4 × 4 (**c**). Peaks obtained in this way are shown with *grey lines*. Reprinted with permission from [80]

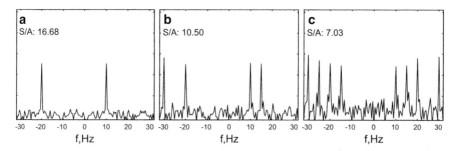

Fig. 14 Decrease in signal-to-artifact ratio with growing number of peaks: (**a**) 2, (**b**) 4, (**c**) 8. S/N was calculated using "true" peak amplitude (not influenced by artifacts)

lower the average signal-to-artifact ratio. This makes nuFT processing more challenging when applied to spectra featuring large numbers of signals with high dynamic range of peak intensities (e.g., NOESY). In this case, artifact-cleaning algorithms may be employed (see Sect. 7).

8.2 Sparse MFT

According to (12) it is possible to choose arbitrarily frequency points for FT, e.g., to calculate just an interesting region(s) of a spectrum. This approach is of particular use when dimensionality and/or resolution is high and the full spectral matrix would be of an extremely large size. There are various possibilities for restricting spectral space to the regions of interest, depending on the type of spectrum and type of information to be extracted [80]. All of them are based on prior examination of other (simpler) spectra. The restriction not only allows one to save disk space, but also accelerates calculations and facilitates data analysis.

8.2.1 "Slice" MFT

In spectra of high dimensionality, peak coordinates in some spectral dimensions are usually known from the spectrum of lower dimensionality (later called "basic spectrum"). The complete and regular frequency grid is not needed in these dimensions and they may be reduced to a set of frequencies corresponding to the tops of peaks [81] (see Fig. 15). The number of lower-dimensional (e.g., 2D) cross-sections obtained with this approach is equal to the number of peaks found in the basic spectrum. Noteworthy, the basic spectrum, used for frequency selection, should also be recorded with high resolution, as an accuracy in determination of peaks frequencies is crucial here. Such a procedure dramatically reduces the amount of data to be stored. In general, the size of data matrix of N-dimensional spectrum is equal to

$$\text{size} = m_1 \cdot m_2 \cdot \ldots \cdot m_N, \qquad (48)$$

where: m_i is a number of spectral points in the ith dimension.

If the frequencies of the first k dimensions are "reduced" during FT, the data matrix size becomes

$$\text{size} = ns \cdot m_{k+1} \cdot m_{k+2} \cdot \ldots \cdot m_N, \qquad (49)$$

where ns is a number of frequency sets obtained from a lower-dimensional spectrum used for Sparse MFT (SMFT). For instance, let us assume that the number of spectral points in each dimension of a 5D data set is equal to 128 and SMFT is

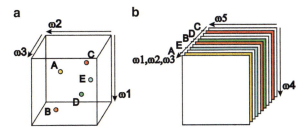

Fig. 15 The idea of a "slice" SMFT. (a) Scheme of a 3D spectrum. Frequency coordinates of peaks from this spectrum (labeled *A–E*) are used as the basis for SMFT calculation. (b) Scheme of a 5D spectrum. Three frequency dimensions ω_1, ω_2, and ω_3, which correspond to nuclei observed in 3D spectrum, are symbolized by one axis; two other dimensions (ω_4 and ω_5) are shown on separate axes. Only 2D (ω_4–ω_5) cross-sections that contain peaks (marked with *colors*) are calculated in SMFT. Reprinted with permission from [80]

performed on the basis of a 3D spectrum containing 150 peaks. In this case, the total size of the resulting set of 2D planes will be $128 \cdot 128 \cdot 128/150 \cong 13981$ times smaller than the size of the full 5D spectrum, which in practice means reduction of the file size from about 100 GB to approximately 10 MB. Moreover, a set of lower-dimensional spectra is easier to handle than one spectrum of high dimensionality (see Sect. 9).

8.2.2 "Cube" MFT

Using techniques of extraordinary resolution, it is possible to measure efficiently peak splitting (E.COSY pattern) associated with internuclear couplings [79]. By increasing maximum evolution times one can reach peak width determined practically only by relaxation rate. However, ultra-narrow peaks require enhanced digital resolution (number of points per Hertz) to be properly visualized. This often causes the need to use another procedure employing reduced frequency space.

Prior to processing of such high-resolution data, positions of peaks should be roughly determined from an equivalent decoupled (i.e., with singlets) spectrum (or spectra) of lower resolution. Afterwards, the spectrum of high resolution is calculated only in the close vicinity of these peaks positions, resulting in a set of full-dimensional "cubes" (see Fig. 16). In each "cube" the numerical resolution should be sufficiently high to visualize the multiplets and determine coupling constants. Again, reduction of the required disk space is significant. The size of data matrix is reduced from that defined by (48) to the following value:

$$\text{size} = ns \cdot m_1 \cdot \frac{sw_1^{loc}}{sw_1} \cdot m_2 \cdot \frac{sw_2^{loc}}{sw_2} \cdot \ldots \cdot m_N \cdot \frac{sw_N^{loc}}{sw_N}, \quad (50)$$

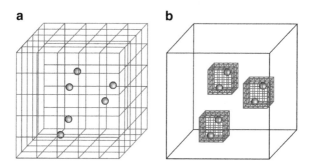

Fig. 16 The idea of a "cube" SMFT. (**a**) Scheme of a full 3D spectrum, containing peaks revealing E.COSY multiplet structure. The digital resolution is too low to approximate properly the narrow components of multiplets. (**b**) Scheme of a set of "cubes", calculated just in vicinities of peaks, featuring much higher digital resolution. Determination of small coupling constants is possible. Reprinted with permission from [80]

where: *ns* is a number of frequency sets used for SMFT, sw_i is spectral width in dimension *i* of full spectrum, and sw_i^{loc} is spectral width in dimension *i* of a single "cube".

For example, in the case of a 4D spectrum, when spectral width of a "cube" is in each dimension ten times smaller than the full spectral width in this dimension, and the number of "cubes" is 150, the data set is reduced about 40,000 times. Typically, it can result in reduction of disk space requirement from tens of terabytes to the order of a gigabyte.

9 Applications

The interpretation of one- and two-dimensional spectra of large biomolecules such as proteins and nucleic acids is usually impossible due to the large number of highly degenerated peaks. Hence, even for the medium-sized molecules, it is necessary to use isotopic enrichment with ^{13}C and ^{15}N nuclei, and to perform triple-resonance 3D NMR experiments for resonance assignment and extraction of structural constraints. However, as we pointed out above, the resolution of conventionally acquired 3D spectra, is limited by sampling requirements. Therefore, it is rarely possible to obtain line widths close to the natural ones in a reasonable time, even for very fast-relaxing molecules. The conventional 4D spectra, such as ^{15}N,^{13}C, or ^{13}C, ^{13}C-edited NOESY experiments, are rarely employed owing to the short evolution times achievable. On the other hand, NMR spectra of biomolecules feature relatively narrow and well-defined spectral regions such as H$_N$, N$_H$, C', Cα, and CαCβ in proteins. This feature allows the development of numerous multidimensional experiments, which correlate spin interactions in different dimensions. Thus, the most important applications of sparse sampling techniques are focused on the important field of structural studies of biomolecules in solution. Sparse non-uniform sampling and FT enable acquisition and processing of multidimensional NMR spectra featuring extraordinary resolution, such as 4-6D NMR spectra dedicated to resonance assignment, techniques for precise determination of coupling constants from 3D and 4D experiments and proton–proton contacts from well-resolved NOESY spectra.

9.1 Development and Implementations

The early applications of sparse sampling and FT processing were devoted to demonstrating the features of the proposed methods rather than to cases of a really demanding nature. Kazimierczuk and coworkers compared 3D HNCO spectra of human ubiquitin employing radial and spiral sampling, showing significant advantages of the latter [44]. Shortly after that, Marion demonstrated FT processing in polar coordinates in application to radially sampled 3D HNCO of human ubiquitin [55]. Next was the work of Coggins and Zhou [56], who formulated the expression for polar FT and applied it for 3D TROSY-HNCO for ^{13}C/^{15}N/^{2}H-labeled OTU protein. In the consecutive works all three groups concentrated on the reduction of the artifact level. Koźmiński's and Marion's groups switched to random sampling, motivating it by the lower intensity and noise-like nature of artifacts in randomly sampled spectra. In the following works, the issue of approximation of Fourier integral was discussed. Kazimierczuk and co-workers [46] demonstrated that the surface integration using Delaunay triangulation improves S/A only for sampling above the Nyquist density. It was also shown that in the case of unweighted FT the S/A ratio does not depend on the relative samples density (see Fig. 8). The usability of the method was verified on 3D HNCA, HNCACB, and ^{15}N-edited NOESY experiments on ubiquitin, using random sampling with exponential and Gaussian distributions of sampling points. Later on, the same authors [60] showed that the simple regularization of the samples distribution reduces artifacts in the signal vicinity. Moreover, it was demonstrated that this effect is more pronounced in comparison with Voronoi tessellation used as an integral quadrature rule. Additionally, in this work the usability of a simple variant of CLEAN algorithm (see Sect. 7) was demonstrated on the 3D ^{15}N-edited NOESY spectrum of ubiquitin. Pannetier and coworkers [64] for the first time applied random sampling and FT processing for intrinsically unstructured protein, namely 60-residue N$_{TAIL}$ (443–501) fragment of nucleoprotein N from the paramyxovirus Sendai. They obtained backbone resonance assignment using two 3D CBCANH and CBCA(CO)NH experiments with 6.5-fold undersampling. At the same time, Coggins and Zhou introduced concentric ring sampling, demonstrating its advantages over the initially used radial alternative, and employed it for 3D HNCO of uniformly ^{13}C, ^{15}N labeled spectrum of the B1 domain of protein G (GB1) [59]. The next development in Zhou's group was concentric shell sampling adjusted to a fine grid which was employed for the 4-D HCCH-TOCSY [83]. In this work artifact suppression was accomplished by an adaptation of the CLEAN algorithm (see Sect. 7).

The influence of different constrained random sampling schedules on Point Spread Function was further investigated by Kazimierczuk and coworkers considering both artifact level and distribution [61]. It was shown that Poisson disk sampling provides the largest low-artifact area in the signal vicinity. The new sampling schemes were verified by application to the 3D HNCACB and ^{15}N-edited NOESY-HSQC acquired for human ubiquitin. The analysis of signal-to-artifact ratio with respect to relative sampling density and dimensionality was analyzed in the next work from the same group [45]. It was proven that for random sampling S/A ratio depends neither on sampling density nor dimensionality of the experiment. These results were experimentally confirmed by acquisition of 5D HC(CC-TOCSY)CONH, performed for doubly labeled human ubiquitin within 0.0054% of the time necessary for analogical conventional experiment.

9.2 Easy Resonance Assignment in Proteins Using the Spectra of High Dimensionality

High resolution and dimensionality achievable in spectra acquired with the use of sparse random sampling and processed by FT feature a significant improvement in peak dispersion. This facilitates resonance frequency assignment, especially in demanding cases such as intrinsically disordered proteins. The first such example, mentioned above, was the backbone assignment of intrinsically unstructured 60-residue N_{TAIL} (443–501) fragment of nucleoprotein N from the paramyxovirus Sendai using the 3D experiments [64].

After the feasibility of 5D experiments acquired by random sampling and SMFT processing by Kazimierczuk et al. was demonstrated [45], the same group proposed a set of 4D (HNCOCA, HNCACO, HNCACACB, HN(CA)NH, and HabCabNH) [81], and later 5D (HN(CA)CONH, HabCabCONH) experiments [80] dedicated to the effective protein backbone signal assignment. All of these techniques employ sparse random sampling and FT processing to achieve high resolution spectra in a tiny fraction of the time needed conventionally. The 4D experiments were tested on two proteins differing in size, i.e., a protein interacting with NIMA-kinase from *Cenarcheaum symbiosum* (96 a.a. residues) and maltose binding protein (371 a.a. residues) (see, for example, Fig. 17).

The feasibility of the 5D techniques was demonstrated using the sample of 5–79 fragment of bovine Ca^{2+}-loaded Calbindin D9K P47M mutant [81]. However, the true test of the new assignment strategy was performed on the particularly demanding case of the δ subunit of RNA polymerase from *Bacillus subtilis* containing a disordered C-terminal region of 81 amino acids with a highly repetitive sequence [92]. While the backbone assignment of this protein appeared to be unachievable using conventional 3D techniques, the strategy based on the new 5D experiments (HN(CA)CONH, HabCabCONH, and HC(CC-TOCSY)CONH) provided a complete backbone and side-chain assignment (see Fig. 18).

9.3 Determination of Coupling Constants in Proteins

Backbone scalar couplings are widely used in NMR studies of structure and dynamics of biomolecules [93]. Additionally, there is a substantial interest in precise determination of residual dipolar couplings for structural studies of weakly oriented biomolecules. Most of the relevant coupling constants in proteins are rather small – of the magnitude from a few to a hundred hertz. Therefore, in order to achieve the sufficient resolution in indirectly measured dimensions, the majority of traditional methods devoted to coupling constants determination in biomolecules are limited to two-dimensional techniques, which frequently suffer from peak overlap. However, the random sampling of evolution time domain allows one to obtain spectra of resolution limited only by transverse relaxation

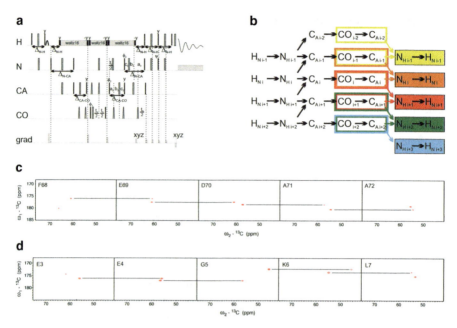

Fig. 17 Example of application of 4D HNCACO technique. (**a**) Pulse sequence. Evolution for CO is in the real-time mode, and for N and CA in semi-constant-time mode ($a_i = (t_i + \Delta)/2$, $b_i = t_i(1 - \Delta/t_{maxi})/2$, $c_i = \Delta(1 - t_i/t_{maxi})/2$) or constant-time mode ($a_i = (\Delta + t_i)/2$, $b_i = 0$, $c_i = (\Delta - t_i)/2$), where Δ stands for Δ_{N-CA} and Δ_{CA-CO}, respectively, t_i is the evolution time in ith dimension and t_{maxi} is the maximal length of evolution time delay. Delays were set as follows: $\Delta_{N-H} = 5.4$ ms $\Delta_{N-CA} = 22$ ms $\Delta_{CA-CO} = 6.8$ ms. (**b**) Coherence transfer in the peptide chain. Amide nitrogen and proton frequencies (*filled colored rectangles*) are fixed during Fourier transformation. Each plane contains CO–CA peak for i and $i-1$ residue. (**c**) 2D spectral planes for CsPin protein obtained by SMFT procedure performed on the 4D HNCACO randomly sampled signal (Poisson disk sampling) with "fixed" H_N and N frequencies obtained from 3D HNCO peak list. (**d**) 2D spectral planes for MBP obtained in the same manner. Reprinted with permission from [81]

and suffices to differentiate multiplet components. Moreover, when couplings with passive spins are resolved in two or more dimensions, the E.COSY [94] multiplet patterns provide valuable information about relative signs of coupling constants. Kazimierczuk and co-workers [79] showed an example of a 3D HNCO-Cα-coupled spectrum of ubiquitin protein. Each peak in this spectrum reveals 3D E.COSY pattern due to couplings with two passive Cα spins. Thus, six coupling constants of H_N, N, and C′ with intra- and inter-residual Cα spins can be determined. The resolution achieved in this experiment would require over a month of conventional acquisition, making it impractical. The coupling constants measured from 3D HNCO-Cα-coupled experiment revealed correlation with φ and ψ protein backbone torsional angles. Later on, a feasibility of determination of a 4D E.COSY patterns was also shown and exemplified with the 4D HNCACO-{Hα} experiment for the sample of 5–79 fragment of bovine Ca^{2+}-loaded Calbindin D9K P47M mutant [80]; see Fig. 19. In this experiment the "cube"-SMFT procedure was employed in order to achieve extraordinary disc space savings.

Generalized Fourier Transform for Non-Uniform Sampled Data   117

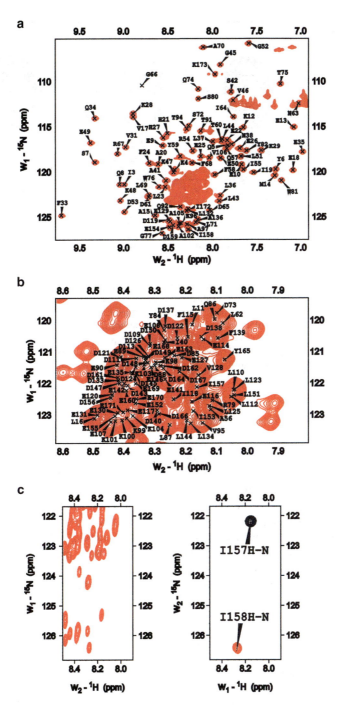

Fig. 18 ^{1}H,^{15}N-HSQC spectra of RNA polymerase δ subunit. (**a**) Entire spectrum. (**b**) With the central region expanded. Central region of ^{1}H,^{15}N-HSQC spectrum (**c**) and the corresponding region of a 2D cross-section extracted from the 5D HN(CA)CONH spectrum (*right*). The 2D

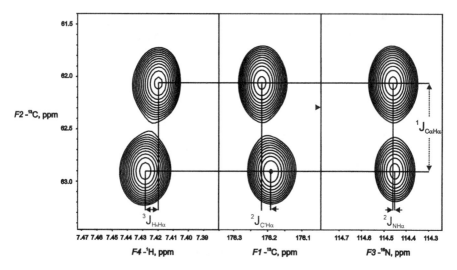

Fig. 19 Experimental example of ultra-high resolution multidimensional NMR spectra obtained by the proposed technique: I74 intra-residual resonance from 89 h 4D HNCACO-{H$_\alpha$}-coupled experiment acquired for 5–79 fragment of bovine Ca^{2+}-loaded Calbindin protein. Depicted cross-sections of 4D "cube" 50 × 450 × 40 × 100 Hz surrounding the peak allow determination of coupling constants from resolved 4D E.COSY pattern. $^1J_{C\alpha H\alpha} = 135.9$ Hz, $^3J_{HNH\alpha} = 5.8$ Hz, $^2J_{C'H\alpha} = -5.0$ Hz, $^2J_{NHH\alpha} = -1.0$ Hz with numerical resolution of 0.4 Hz/point, 1.7 Hz/point, 0.2 Hz/point, and 0.7 Hz/point in dimensions F$_1$, F$_2$, F$_3$, and F$_4$, respectively. Reprinted with permission from [80]

9.4 Heteronuclear-Edited NOESY Experiments

NOESY experiments are still the primary source of structural information. The presence of the cross-peaks in NOESY spectra indicates spatial proximity of nuclei, and their integral is proportional to the r^{-6}, where r denotes internuclear distance. However, NOESY spectra are significantly more difficult to obtain in comparison with other NMR techniques. The most important differences are the large number of correlation peaks, dependent on the number of interacting proton nuclei, and a high dynamic range of peak amplitudes up to two to three orders of magnitude. Consequently, NOESY spectra require an excellent sensitivity and almost perfect suppression of spectral artifacts. Moreover, in order to preserve the relationship between peak integral and internuclear distances, the linearity of the method should

---

Fig. 18 (continued) cross-section was obtained by fixing frequencies in ω_3, ω_4, and ω_5 dimensions to the values of chemical shifts of ^{13}C' of I157, ^{15}N of I158, and ^{1}H$_N$ of I158, respectively. Peaks corresponding to the sequential and intraresidual correlations are displayed in *black* and *red*, respectively. Experiment was acquired within 20 h on 700 MHz spectrometer using RT-probe. Only a fraction of 0.00034% points was collected in indirect time domains. See [92] for further details

be maintained. Thus, this type of applications is very demanding for all sparse NMR techniques. In the case of nuFT processing, effective artifact suppression is necessary.

Kazimierczuk and co-workers applied their semi-automatic CLEAN procedure to suppress artifacts in a randomly sampled ^{15}N-labeled NOESY-HSQC spectrum of ubiquitin [60]. It was demonstrated that the process does not systematically influence relative peak amplitudes, and is therefore applicable to NOESY spectra. Similar conclusions were later drawn by Stanek and Koźmiński [85], and by Werner-Allen and co-workers [84], who compared their reconstructions with conventionally sampled three-dimensional spectra of the same spectral resolution. The algorithm proposed by Kazimierczuk and co-workers was later also applied to higher-dimensional experiments [80].

Coggins and Zhou implemented the CLEAN algorithm to process four-dimensional spectra [83], with only slight modifications with respect to the original procedure from radio astronomy. The advantages of CLEAN processing in conjunction with Randomized Concentric Shell Sampling were demonstrated on the 4D HCCH-TOCSY spectrum of 56 a.a. GB1 protein. In this experiment, 1.2% of samples was used, and CLEAN was shown to decrease the apparent noise from 2.4 to 1.4 of thermal noise level on average.

The same program was later used to process the 4D amide–amide diagonal-suppressed TROSY-NOESY-TROSY (ds-TNT) spectrum of 23 kDa C13S Sssu72 protein [84]. The largest decrease in apparent noise level due to the CLEAN process was 22%. The application of sparse sampling and FFT-CLEAN processing allowed a more than tenfold reduction in experimental time in comparison with the conventional approach to acquisition. The experiment was shown to provide valuable information on distance restraints between amide protons by avoiding the ambiguities and frequent resonance overlap typical for 3D NOESY spectra of large proteins (see Fig. 20).

A more challenging example was demonstrated by Stanek and Koźmiński [85], who applied their algorithm to 3D ^{15}N- and ^{13}C-labeled NOESY spectra of ubiquitin *without* suppression of diagonal peaks. The efficiency of artifact suppression was investigated by comparison of the reconstruction with the conventionally acquired reference spectrum. Less than 2% of peaks were missing, and about 1.5% false peaks were reported. The correlation coefficient between peak volumes of $R^2 = 0.998$ was obtained.

9.5 3D Spectra of Complex Organic Compounds

Compared to the progress and the variety of new multidimensional methods proposed in the area of biomolecules, in the field of organic molecules the development is slower, caused mainly by less demanding applications and additional experimental limitations. However, in the case of complex organic molecules, it is sometimes necessary to add the third dimension to separate crowded, overlapping

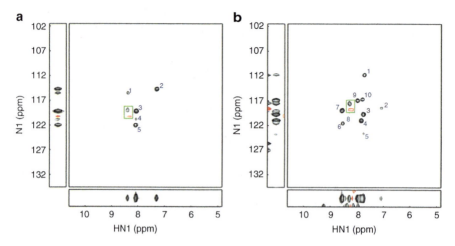

Fig. 20 $F_1(H_N)/F_2(N)$ cross-sections from 4D amide-amide ds-TNT spectrum of C13S Ssu72 protein. Residual diagonal peaks of Ile176 (**a**) and overlapped Leu72 and Asn92 (**b**) are enclosed with *green boxes*. Corresponding strips in each panel, plotted from conventionally sampled 3D ds-TNT spectra, show severe overlap of the amide-amide cross-peaks. In contrast, in the sparsely sampled 4D ds-TNT spectrum the peaks were clearly resolved and assigned. Reprinted with permission from [84]

signals in order to avoid ambiguities in the spectral assignment. The use of unconventional approaches in the acquisition of multidimensional NMR signals makes it possible to record 3D NMR spectra of small molecules in shorter experimental time. So far there are only very few examples of using 3D NMR experiments dedicated to the spectral assignment [95–99] and the measurement of coupling constants [100].

Generally, in organic chemistry two-dimensional spectra are widely used, while 3D NMR spectra of small molecules were hardly achievable, because of the very long measurement time required in the conventional approach. In contrast to proteins, organic compounds at the natural isotopic abundance are more demanding due to the low sensitivity and the necessity of sampling the wide frequency range especially in the ^{13}C dimension. On the other hand, the slower transverse relaxation rates allow one to achieve narrow peaks, which again is limited by sampling. That is why in many cases these problems precluded the full assignment of NMR signals and the evaluation of coupling constants of organic compounds. Due to the employment of non-uniform sampling, the application of multidimensional NMR spectra in the structure investigation of organic molecules became practically possible.

Recently the method employing DNP for recording heteronuclear 2D NMR spectra of small drug-like molecules was proposed by Ludwig and coworkers [101]. This method ensures significant improvement in sensitivity due to the high spin polarization, but limits the number of points sampled in indirectly detected dimension, so the combination with the non-uniform sampling scheme was necessary.

Three-dimensional NMR spectra based on random sampling of the evolution time space followed by MFT processing were successfully applied by Misiak and Koźmiński in the structural analysis of complex organic compounds [95]. Three new 3D NMR techniques (TOCSY-HSQC, COSY-HMBC, and HSQMBC) which allow the spectral assignment have been proposed. The comparison of 3D spectra of strychnine recorded in the conventional way with that acquired using randomly distributed data points in the evolution time space revealed that by using this new approach it is possible to acquire 3D spectra in reasonable experimental time, while retaining high resolution in indirectly detected domains (see Fig. 21). The use of 3D TOCSY-HSQC and 3D COSY-HMBC allowed for the complete assignment of ^{1}H and ^{13}C chemical shifts of natural abundance prenol-10 [96], which was earlier impossible by employing 1D and 2D spectra, mostly because of the signal overlapping caused by similarity of the ten isoprene units. The application of 3D HSQC-TOCSY spectra with E.COSY- type multiplets enabled the accurate determination of heteronuclear coupling constants of organic molecules in an overnight experiment [100].

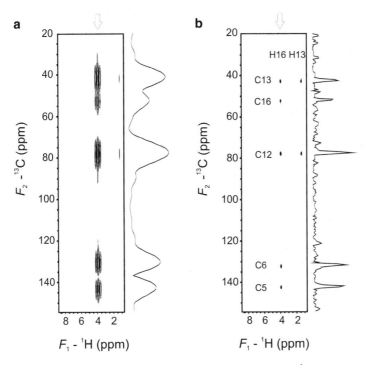

Fig. 21 Comparison of 3D COSY-HMBC F_1/F_2 cross-sections for F_3 (^{1}H) = 3.904 ppm, i.e., resonance frequency of the H16 atom of the strychnine molecule: (**a**) conventional and (**b**) random sampling of t_1/t_2 evolution time space. The spectra were recorded in the same experimental time, and transformed with the resolution of 128 × 256 × 1,024 points in F_1, F_2, and F_3, respectively. The *vertical arrows* indicate the positions of the extracted traces. Reprinted with permission from [95]

In the case of naturally abundant complex organic compounds playing important biological roles, detailed structural analysis is very important. We believe that, in the future, recording of sparsely sampled 3D NMR spectra should also become a routine procedure for the structural analysis of complex organic molecules.

10 Conclusions

The application of sparse sampling for the acquisition of multidimensional NMR spectra leads to the presence of spectral artifacts. They appear in a regular form (e.g., ridges, rings) for the regular sampling, and resemble the noise in the case of random sampling. The spectral reconstruction aims to obtain the spectrum with minimized artifact level. Among a variety of reconstruction methods, the FT has favorable computational requirements. The important feature of off-grid random sampling and FT processing is the independence of artifact intensity of the degree of sparseness, and decreasing of artifacts with the square root of number of sampled points. Therefore, it should not be applied for the acceleration of experiments attributing conventional resolution, which is a usual task of "fast NMR" techniques. FT is rather the method of choice for the acquisition and processing of spectra of high dimensionality (4D–6D) or of high resolution, approaching natural line width. Frequently, the artifact level in such spectra is low enough to allow their interpretation without further processing. However, for the analysis of high dynamic range spectra featuring a large number of signals, such as, for example, NOESY experiments, additional artifact "cleaning" is required. Until now, the number of such applications is still minor. However, we expect that it will grow systematically in parallel with dissemination of the necessary software. We believe that the random sparse sampling and FT processing could be aimed at a variety of new applications, especially in the field of NMR-based structural studies of biomolecules.

Acknowledgments This work has been supported in part by the EC contract EAST-NMR no 228461. Bio-NMR project under the seventh Framework Programme of the EC grant agreement 261863 for conducting the research is gratefully acknowledged. A.Z.-K. thanks the Foundation for Polish Science for supporting her with the MPD Programme that was co-financed by the EU European Regional Development Fund. K.K. thanks the Foundation for Polish Science for supporting him with the KOLUMB scholarship. M.M. and W.K. acknowledge the Ministry of Science and Higher Education of Poland for the grant N204 137937 for the years 2009–2010. We are grateful to V. Motáčková, J. Nováček, L. Žídek, H. Šanderová, L. Krásný, and V. Sklenář for providing Fig. 18.

References

1. Ernst RR, Anderson WA (1966) Rev Sci Instrum 37:93
2. Jeener J (1971) Basko Polje, Yugoslavia
3. Aue WP, Bartholdi E, Ernst RR (1976) J Chem Phys 64:2229
4. Wüthrich K (1986) NMR of proteins and nucleic acids. Wiley, New York

5. Montelione GT, Wagner G (1989) J Am Chem Soc 111:5474
6. Ikura M, Kay LE, Bax A (1990) Biochemistry 29:4659
7. Bax A, Grzesiek S (1993) Acc Chem Res 26:131
8. Yamazaki T, Lee W, Arrowsmith CH, Muhandiram DR, Kay LE (1994) J Am Chem Soc 116:11655
9. Sattler M, Schleucher J, Griesinger C (1999) Prog Nucl Magn Reson Spectrosc 34:93
10. Moskau D (2002) Concepts Magn Reson 15:164
11. Szantay C (2008) Concepts Magn Reson Part A 32A:373
12. Schanda P, Kupče Ē, Brutscher B (2005) J Biomol NMR 33:199
13. Schanda P, Van Melckebeke H, Brutscher B (2006) J Am Chem Soc 128:9042
14. Lescop E, Schanda P, Brutscher B (2007) J Magn Reson 187:163
15. Frydman L, Scherf T, Lupulescu A (2002) Proc Natl Acad Sci USA 99:15858
16. Mishkovsky M, Frydman L (2008) Chemphyschem 9:2340
17. Mishkovsky M, Kupče Ē, Frydman L (2007) J Chem Phys 127:034507
18. Gal M, Frydman L (2010) J Magn Reson 203:311
19. Mishkovsky M, Frydman L (2009) Annu Rev Phys Chem 60:429
20. Tal A, Frydman L (2010) Prog Nucl Magn Reson Spectrosc 57:241
21. Led JJ, Gesmar H (2010) In: Morris GA, Emsley JW (eds) Multidimensional NMR methods for the solution state. Wiley, Chichester, p 131
22. Armstrong GS, Cano KE, Mandelshtam VA, Shaka AJ, Bendiak B (2004) J Magn Reson 170:156
23. Armstrong GS, Mandelshtam VA, Shaka AJ, Bendiak B (2005) J Magn Reson 173:160
24. Meng X, Nguyen BD, Ridge C, Shaka AJ (2009) J Magn Reson 196:12
25. Brüschweiler R (2004) J Chem Phys 121:409
26. Brüschweiler R, Zhang FL (2004) J Chem Phys 120:5253
27. Snyder DA, Bruschweiler R (2009) J Phys Chem A 113:12898
28. Snyder DA, Xu Y, Yang D, Bruschweiler R (2007) J Am Chem Soc 129:14126
29. Trbovic N, Smirnov S, Zhang F, Bruschweiler R (2004) J Magn Reson 171:277
30. Zhang FL, Bruschweiler R (2004) J Am Chem Soc 126:13180
31. Robin M, Delsuc M-A, Guittet E, Lallemand J-Y (1991) J Magn Reson 92:645
32. Jeannerat D (2007) J Magn Reson 186:112
33. Coggins BE, Venters RA, Zhou P (2010) Prog Nucl Magn Reson Spectrosc 57:381
34. Kazimierczuk K, Stanek J, Zawadzka-Kazimierczuk A, Koźmiński W (2010) Prog Nucl Magn Reson Spectrosc 57:420
35. Kim S, Szyperski T (2003) J Am Chem Soc 125:1385
36. Ding KY, Gronenborn AM (2002) J Magn Reson 156:262
37. Koźmiński W, Zhukov I (2003) J Biomol NMR 26:157
38. Hiller S, Fiorito F, Wüthrich K, Wider G (2005) Proc Natl Acad Sci USA 102:10876
39. Malmodin D, Billeter M (2005) J Magn Reson 176:47
40. Kupče Ē, Freeman R (2005) J Magn Reson 173:317
41. Kupče Ē, Freeman R (2003) J Am Chem Soc 125:13958
42. Kupče Ē, Freeman R (2004) Concepts Magn Reson Part A 22A:4
43. Kupče Ē, Freeman R (2004) J Biomol NMR 28:391
44. Kazimierczuk K, Koźmiński W, Zhukov I (2006) J Magn Reson 179:323
45. Kazimierczuk K, Zawadzka A, Koźmiński W (2009) J Magn Reson 197:219
46. Kazimierczuk K, Zawadzka A, Koźmiński W, Zhukov I (2006) J Biomol NMR 36:157
47. Laue ED, Skilling J, Staunton J, Sibisi S, Brereton RG (1985) J Magn Reson 62:437
48. Hoch JC, Stern AS (1996) NMR data processing. Wiley-Interscience, New York
49. Mobli M, Hoch JC (2008) Concepts Magn Reson Part A 32A:436
50. Luan T, Jaravine V, Yee A, Arrowsmith CH, Orekhov VY (2005) J Biomol NMR 33:1
51. Jaravine V, Ibraghimov I, Orekhov VY (2006) Nat Meth 3:605
52. Jaravine VA, Zhuravleva AV, Permi P, Ibraghimov I, Orekhov VY (2008) J Am Chem Soc 130:3927

53. Szantay C (2008) Concepts Magn Reson Part A 32A:1
54. Nyquist H (2002) Proc IEEE 90:280
55. Marion D (2006) J Biomol NMR 36:45
56. Coggins BE, Zhou P (2006) J Magn Reson 182:84
57. Szyperski T, Yeh DC, Sukumaran DK, Moseley HN, Montelione GT (2002) Proc Natl Acad Sci USA 99:8009
58. Malmodin D, Billeter M (2006) Magn Reson Chem 44:185
59. Coggins BE, Zhou P (2007) J Magn Reson 184:207
60. Kazimierczuk K, Zawadzka A, Koźmiński W, Zhukov I (2007) J Magn Reson 188:344
61. Kazimierczuk K, Zawadzka A, Koźmiński W (2008) J Magn Reson 192:123
62. Hyberts SG, Takeuchi K, Wagner G (2010) J Am Chem Soc 132:2145
63. Tarczynski A, Allay N (2003) 7th world multiconference on systemics, Cybernetics and Informatics, Proceedings, Vol IV, p 344
64. Pannetier N, Houben K, Blanchard L, Marion D (2007) J Magn Reson 186:142
65. Press WH, Teukolsky SA, Vetterling WT, Flannery BP (2007) Numerical recipes: the art of scientific computing. Cambridge University Press, Cambridge/New York
66. Davis PJ, Rabinowitz P (1984) Methods of numerical integration. Academic, New York
67. Barthold E, Ernst RR (1973) J Magn Reson 11:9
68. Hoch JC, Stern AS, Wagner G (1995) J Cell Biochem, Suppl. 21B: 76
69. Matsuki Y, Eddy MT, Herzfeld J (2009) J Am Chem Soc 131:4648
70. Donoho DL (2006) Commun Pure Appl Math 59:797
71. Lustig M, Donoho D, Pauly JM (2007) Magn Reson Med 58:1182
72. Lustig M, Donoho DL, Santos JM, Pauly JM (2008) IEEE Signal Process Mag 25:72
73. Drori I (2007) Eurasip J Adv Signal Process doi:10.1155/2007/20248
74. Shrot Y, Frydman L (2011) J Magn Reson 209:352
75. Hyberts SG, Frueh DP, Arthanari H, Wagner G (2009) J Biomol NMR 45:283
76. Stern AS, Donoho DL, Hoch JC (2007) J Magn Reson 188:295
77. Marion D (2005) J Biomol NMR 32:141
78. Jiang B, Jiang X, Xiao N, Zhang X, Jiang L, Mao XA, Liu M (2010) J Magn Reson 204:165
79. Kazimierczuk K, Zawadzka A, Koźmiński W, Zhukov I (2008) J Am Chem Soc 130:5404
80. Kazimierczuk K, Zawadzka-Kazimierczuk A, Koźmiński W (2010) J Magn Reson 205:286
81. Zawadzka-Kazimierczuk A, Kazimierczuk K, Koźmiński W (2010) J Magn Reson 202:109
82. Coggins BE, Venters RA, Zhou P (2004) J Am Chem Soc 126:1000
83. Coggins BE, Zhou P (2008) J Biomol NMR 42:225
84. Werner-Allen JW, Coggins BE, Zhou P (2010) J Magn Reson 204:173
85. Stanek J, Koźmiński W (2010) J Biomol NMR 47:65
86. Högbom JA (1974) Astron Astrophys Suppl Ser 15:417
87. Keeler J (1984) J Magn Reson 56:463
88. Shaka AJ, Keeler J, Freeman R (1984) J Magn Reson 56:294
89. Barna JCJ, Tan SM, Laue ED (1988) J Magn Reson 78:327
90. Barna JCJ, Laue ED, Mayger MR, Skilling J, Worrall SJP (1987) J Magn Reson 73:69
91. Davies SJ, Bauer C, Hore PJ, Freeman R (1988) J Magn Reson 76:476
92. Motáčková V, Nováček J, Zawadzka-Kazimierczuk A, Kazimierczuk K, Žídek L, Šanderová H, Krásný L, Koźmiński W, Sklenář V (2010) J Biomol NMR 48:169
93. Eberstadt M, Gemmecker G, Mierke DF, Kessler H (1995) Angew Chem Int Ed Engl 34:1671
94. Griesinger C, Sorensen OW, Ernst RR (1985) J Am Chem Soc 107:6394
95. Misiak M, Koźmiński W (2007) Magn Reson Chem 45:171
96. Misiak M, Koźmiński W, Kwasiborska M, Wójcik J, Ciepichal E, Swiezewska E (2009) Magn Reson Chem 47:825
97. Meier S, Benie AJ, Duus JO, Sorensen OW (2009) J Magn Reson 200:340
98. Meier S, Petersen BO, Duus JO, Sorensen OW (2009) Carbohydr Res 344:2274
99. Findeisen M, Bermel W, Berger S (2006) Magn Reson Chem 44:455
100. Misiak M, Koźmiński W (2009) Magn Reson Chem 47:205
101. Ludwig C, Marin-Montesinos I, Saunders MG, Gunther UL (2010) J Am Chem Soc 132:2508

Applications of Non-Uniform Sampling and Processing

Sven G. Hyberts, Haribabu Arthanari, and Gerhard Wagner

Abstract Modern high-field NMR instruments provide unprecedented resolution. To make use of the resolving power in multidimensional NMR experiment standard linear sampling through the indirect dimensions to the maximum optimal evolution times (~ 1.2 T_2) is not practical because it would require extremely long measurement times. Thus, alternative sampling methods have been proposed during the past 20 years. Originally, random nonlinear sampling with an exponentially decreasing sampling density was suggested, and data were transformed with a maximum entropy algorithm (Barna et al., J Magn Reson 73:69–77, 1987). Numerous other procedures have been proposed in the meantime. It has become obvious that the quality of spectra depends crucially on the sampling schedules and the algorithms of data reconstruction. Here we use the forward maximum entropy (FM) reconstruction method to evaluate several alternate sampling schedules. At the current stage, multidimensional NMR spectra that do not have a serious dynamic range problem, such as triple resonance experiments used for sequential assignments, are readily recorded and faithfully reconstructed using non-uniform sampling. Thus, these experiments can all be recorded non-uniformly to utilize the power of modern instruments. On the other hand, for spectra with a large dynamic range, such as 3D and 4D NOESYs, choosing optimal sampling schedules and the best reconstruction method is crucial if one wants to recover very weak peaks. Thus, this chapter is focused on selecting the best sampling schedules and processing methods for high-dynamic range spectra.

Keywords FM reconstruction · NMR · NOESY · Non-uniform sampling · Triple resonance

S.G. Hyberts, H. Arthanari, and G. Wagner (✉)
Department of Biological Chemistry and Molecular Pharmacology, Harvard Medical School, 240 Longwood Avenue, Boston, MA 02115, USA
e-mail: Gerhard_Wagner@hms.harvard.edu

Contents

1. Introduction 126
2. The Forward Maximum Entropy Reconstruction Relative to Other Procedures 127
3. Description of the Software and Computer Hardware 128
4. Principles of Non-Uniform Sampling 130
5. The Point-Spread Function for Evaluation of Sampling Schedules 131
6. Evaluating Sampling Schedules with the Point-Spread Function 131
7. Performance of Sampling Schedules on a Set of Four Peaks Using FM Reconstruction 134
8. Fourier Transformation of NUS Data Without Reconstruction of Missing Points ... 137
9. Evaluation with Noise, Good and Bad 137
10. Implementing NUS Schedules in Two Indirect Dimensions 140
11. Comparison of Sampling Schedules in Two Dimensions as Used in 3D NMR Spectra 141
12. Comparison of NUS in a 3D NOESY Spectrum 143
13. Discussion 145
References 147

1 Introduction

The introduction of pulsed NMR and Fourier transformation of the time domain data has revolutionized NMR spectroscopy [1]. The routine application of this technology became possible with the fast Fourier transformation (FFT) algorithm [2]. It requires time domain data to be sampled in linear increments to enable its application. This technology has dominated NMR spectroscopy ever since. The sampling procedures used are called linear or uniform sampling. With the availability of higher field magnets and considering the low sensitivity of biological samples new ideas have come up to depart from uniform sampling and use new processing methods to enhance the capabilities of NMR spectroscopy, in particular of biological macromolecules.

The first proposal to abandon linear sampling was made by Barna et al. who suggested placing the sampling points in the indirect dimension with exponentially decreasing separation and transforming the spectra with a maximum entropy algorithm [3] developed by Skilling et al. [4]. This proposal has found many followers since and numerous sampling methods and data processing procedures have been proposed. The approach has been further developed with the Maximum Entropy (MaxEnt) reconstruction tool using a different algorithm [5], and many applications and implementations have followed [6–14]. The principle advantages of non-uniform sampling are increasingly recognized [15]. Besides Maximum Entropy reconstruction, other methods are used for processing non-uniformly recorded spectra, such as the Maximum Likelihood Method (MLM) [16], a Fourier transformation of non-uniformly spaced data using the Dutt–Rokhlin algorithm [17], and multi-dimensional decomposition (MDD) [18–22]. Several other methods have been presented to allow for a rapid acquisition of NMR spectra with suitable processing tools, including radial sampling and GFT [23–31].

While there is still much research to be done to find optimal non-uniform sampling schedules and processing methods, the large benefits are already quite

Applications of Non-Uniform Sampling and Processing

obvious. Nevertheless, most laboratories still acquire multidimensional NMR data with the traditional uniform acquisition schedules and process data with the FFT algorithm. This is despite the fact that linear sampling of 3D and 4D NMR data at modern high-field NMR spectrometers can only cover a fraction of the indirect time domains, and the resolution power of the new instruments is often largely wasted.

Here we analyze the benefits of non-uniform sampling for NMR spectroscopy on challenging biological macromolecules. The application of NUS has been well established with spectra that have no dynamic range problem, such as triple resonance experiments as described previously [6, 32]. This allows recording of multidimensional triple resonance spectra at resolutions matching the resolving power of modern high-field instruments in a reasonable time. The current challenge of NUS approaches is more for experiments with high dynamic range problems, such as 3D and 4D NOESY spectra. Thus, the following is more concerned with this aspect. The following steps are taken in this chapter. First we discuss procedures towards finding optimal sampling methods using point spread functions. Second, we analyze the variation of performance due to the selection of the seed numbers used for creating random sampling schedules in different densities. Third, we compare different sampling schedules when using just straight FFT for reconstruction. Fourth, we analyze the variation of the performance depending on the stochastic type of noise. Fifth and finally, we compare the use of different sampling schedules on an experimental 3D NOESY spectrum. Since the benefits of NUS depend on the processing methods we start out with a discussion of several processing principles but compare different sampling strategies primarily with the FM reconstruction procedure developed in our laboratory [33].

2 The Forward Maximum Entropy Reconstruction Relative to Other Procedures

The purpose of this chapter is to compare different sampling schedules. We do this using the FM reconstruction procedure described in detail previously [32, 33]. In short, FM reconstruction is a minimization procedure that obtains the best spectrum consistent with the non-uniformly sampled time domain data set by minimizing a target function $Q(\mathbf{f})$, which is a norm of the frequency spectrum, such as the Shannon entropy S, or simply the sum of the absolute values of all spectral points. Initially, all data points not recorded are set to zero. This initial time domain data set is Fourier transformed yielding the initial $Q(\mathbf{f})$. Each of the time domain data points that were not recorded are then altered somewhat and transformed independently to create a whole set of somewhat perturbed spectra and hence $Q(\mathbf{f} + \delta_i)$ for each of the altered time domain data points. The relation between $Q(\mathbf{f})$ and $Q(\mathbf{f} + \delta_i)$ for each i defines the gradient ∇, and a Polak Riviere conjugate gradient minimization of the target function $Q(\mathbf{f})$ with respect to the values of the missing time domain data points is carried out. This procedure minimizes $Q(\mathbf{f})$ of all frequency-domain data points while only altering the non-measured time domain data points.

The only variable in this procedure is the choice of the target function; we prefer to use the sum of the absolute values of the frequency domain data points, and the number of iterations to be carried out. Below we use this FM reconstruction procedure for comparing different sampling schedules.

3 Description of the Software and Computer Hardware

The FM reconstruction program has been described previously [32] but some key aspects and recent developments are summarized here. The program requires a gradient for the minimization algorithm. Since the points that vary are in the time domain, and the target function is defined in the frequency-domain, each calculation of a partial derivative requires a forward Fourier Transform of the deviation in the time domain to the frequency-domain. A couple of considerations are immediately possible: first, contemplating that FFT is linear, i.e., FFT($\mathbf{t}_1 + \mathbf{t}_2$) = FFT($\mathbf{t}_1$) + FFT($\mathbf{t}_2$), the set $\mathbf{t}_1$ corresponding to the present set of values of the time domain data and the set $\mathbf{t}_2$ corresponding to a small value at the particular point i to which the partial derivative address, zero elsewhere; the FFT for the present set $\mathbf{t}_1$ and the alteration set can be made separately. Hence, the FFT for the present set $\mathbf{t}_1$ can be made once for all partial derivatives in the gradient. Second, as the set only describes one point, an FFT is not required since the transform is already known as a sinusoidal wave in the frequency domain. Third, if it is possible to store the result in memory for all indices of the gradient, this will also reduce the execution time to a certain extent.

The execution time for each partial derivative of the gradient with the above considerations will be of $O(n_{\max})$, where $n_{\max}$ is the number of points in the trial spectrum. This is due to the fact that a summation of all altered data points in the frequency domain is made for the target function $Q(\mathbf{f})$. With n_{variable} data points in the time domain, the execution time of one conjugate gradient iteration is hence proportional to $n_{\text{variable}} \times n_{\max}$. It follows that the time for reconstruction will be quadratic in n, or $O(n^2)$, depending both on the number of points to be reconstructed and on the size of the reconstructed data. The total time for a complete reconstruction is hence further proportional to the number of iterations ($n_{\text{iterations}}$) of the conjugate gradient iteration and proportional to the number of points kept of the processed uniformly sampled dimension(s) ($n_{\text{processed}}$). The latter usually refers to the kept points after processing the direct dimension and extracting the area of interest, but it may also include the result after processing an indirect dimension that was not obtained non-uniformly.

Except for making the calculation of the gradient as efficient as possible, two approaches for parallelization are possible: (1) farming of the individual calculations for the points kept of the processed uniformly sampled dimension(s)

($n_{\text{processed}}$) and (2) internal parallelization of each of the partial derivatives ($n_{\text{variable}} \times n_{\text{max}}$).

We constructed a program termed mpiPipe, to farm the individual calculations according to (1) above. It does the following. (1) Initiating and connecting with the other processing nodes. (2) Receiving data according to NMRPipe specifications. (3) Once initiating is done, the head node engages each external processor with a job; (a) a task identifier is sent to the external processor, (b) a static command operation is sent to the processor, (c) a unique job order is assigned and kept, allowing asynchronous work flow, (d) the data are prepared and sent, (e) a non-blocking receive is requested. (4) Once a processor node has completed its task, the head node receives it and new data are delegated. (5) Once all processed data have been received from the processing nodes, the processed data are moved from the internal storage to the output pipe according to NMRPipe specifications. Notably, the mpiPipe program may be used for most types of NMRPipe processing on a cluster or a farm via MPI (Message Passing Interface).

As an alternative to using MPI, simple queuing to a cluster is also actively used. As NMRPipe is built on a 2D principle, we are typically using MPI for one-dimensional reconstruction of, e.g., 2D HSQC spectra, whereas the more approachable use of a queuing system comes into question for multi-dimensional reconstructions, such as of triple resonance and four-dimensional spectra. Presently, we are using a 32-node cluster, each equipped with 2 dual 3.0 GHz core Xeon processors with a total of 128 cores.

In order to approach an internal parallelization, specific hardware is required. Fortunately, recent developments within the field of HPC (High Performance Computation) have led to the option of programming GPUs with high numbers of cores. Using CUDA C language, we re-wrote the appropriate calculation of the gradient. Applying the code on a NVIDIA 240 core Tesla C-1060 computer, we find that we can get a speedup of a factor 90 compared to one core of a 3.0 Xeon processor. Practically, this means using a workstation equipped with 4 of these cards (a total of 960 cores), we have achieve a speedup of a factor of 3 relative to our 128-core cluster. Presently, the 448 core Tesla C-2050 has been released, where each card is three times faster than its predecessor. Equipping a workstation with four of these cards hence yielded a single node computer that is nearly ten times faster than our 128-cpu cluster at a fraction (around one tenth) of its cost.

As an alternative to using the high-end of these workstations and graphics cards, a smaller work station can be employed that is capable of using a graphics card comparable to the C-2050, such as the GTX-460, at a cost of less than $1,000. This would yield a speed comparable or faster than our 128-cpu cluster. When working with Open CL, it is also possible to use the ATI graphics cards. We have found that this yields a performance comparable to the NVIDIA C-1060 card.

4 Principles of Non-Uniform Sampling

It is worth noting that one could say that a form of non-uniform sampling was applied prior to the introduction of Fourier transform NMR in routine continuous wave spectroscopy when only the interesting parts of the spectra were scanned. With the introduction of pulsed NMR a set of uniformly sampled equidistant data points spaced by the dwell-time were to be acquired for proper input to the Fast Fourier Transform (FFT) algorithm. This was soon amended with zero filling, which can be considered a new general form of non-uniform sampling.

With the introduction of 2D and generally nD NMR, the number of increments grows exponentially with $t^{(n-1)}$ whereas the information content remains restricted. This means that higher dimensional spectra contain a larger fraction of empty frequency-domain areas with only noise to spread out the signals. While this allows for better identification of signals due to reduced overlap, it is paid for by the need for recording more increments and thus longer measuring times.

The desire for reducing the experimental time has revoked the idea of non-uniform sampling. Is it possible to design a sampling strategy, which does not necessarily inherit the requirement of equidistant data points, in order to reduce the total acquisition time and maintain the ability to distinguish otherwise overlapped signals? If this is the case, can one re-use the saved time and acquire more scans for the measured increments in order to improve sensitivity while maintaining resolution?

As indicated before, the use of FFT requires uniformly sampled equidistant data points. If it is desired to preserve a traditional looking spectrum, the unobtained data points have hence to either be left zero (as with zero filling), or the values of the data points have to be emulated or reconstructed prior to Fourier transformation.

A non-trivial issue with implementing non-uniform sampling is that the traditional test functions for the Fourier Transform (i.e., the set of sinusoidal functions), or even a subset of them, are no longer orthogonal. This leads to artifacts via a mechanism called signal leaking. This does in fact occur for uniform sampling as well: when a signal's frequency is not one of the sinusoidal test-functions, signal leakage is manifested in so-called sinc-wiggles. Since sinusoidal functions no longer provide an orthogonal set when non-uniform sampling is used, signal leaking occurs even when the acquired signal is one of the sinusoidal functions traditionally displayed in a spectrum. In the frequency domain, these leakage artifacts are viewed as the so-called point-spread function. They can easily overshadow the noise, especially when strong signals are present. This is particularly serious in NMR spectra with a large dynamic range. It is hence imperative for non-uniform sampling methods to use schedules where the set of test functions deviates least from being orthogonal.

5 The Point-Spread Function for Evaluation of Sampling Schedules

Intrinsic to all non-uniform sampling is the selection of a sampling schedule. Choosing an optimal schedule is central to the faithful reconstruction of the true spectra from NUS data. Randomly selecting 256 out of 1,024 points can be done in $\binom{1024}{256}$ ways, or more than 10^{248} combinations. A more modest selection, 64 out of 256 data points, still yields 10^{61} combinations. Theoretically, sampling through all of these possibilities would give us some schemes that were obviously not to be considered, such as selecting just the first 256 data points, or picking every fourth point. Here we use random or weighted random selection of sampling points, relying on a random number generator, such as the UNIX *drand48* or equivalent, which only promises a randomization to 2^{48} ways, or just above 10^{14} combinations. Each of these alternative schedules would be generated with a unique seed value. Note that, as we are utilizing a pseudo-random generator, the sequence of random numbers is completely determined by its seed value.

To predict the performance of a sampling schedule we rely on the point spread function. We create a synthetic exponentially decaying signal, select a subset of points with a sampling schedule, and create a spectrum with the FM reconstruction algorithm. The spectrum is then compared with that obtained by FFT from the full uniformly sampled free induction decay, and the difference is expressed in terms of an L^2 norm. This is analogous to a χ^2 analysis, but the values are not normalized. As we have reported previously, the performance of the sampling schedule (lowest L^2 norm) depends on the value of the initial random seed number [34]. This is shown in Fig. 1, where we evaluate different sampling schedules with the point-spread function using 100 different seed numbers. The resulting L^2 differences are sorted from worst (left) to best (right). Here we select 256 out of 1,024 time domain data points, a 25% sampling schedule.

6 Evaluating Sampling Schedules with the Point-Spread Function

We first explore some typical sampling schedules that have been proposed in the literature.

1. Uniformly random

Here 256 out of 1,024 sampling points are selected randomly using the Unix random number generator with different seed numbers, but uniformly spread over the time axis. Figure 1 shows that the deviation from the correct spectrum (L^2 norm) ranges from 95 (worst) to 32 (best). Thus, there is a large variation of fidelity depending on the pick of the seed number.

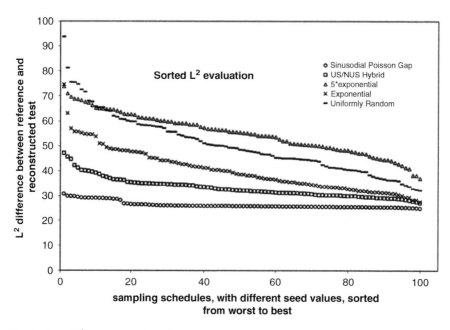

Fig. 1 Sorted L^2 norm evaluation of five sampling strategies for non-uniform sampling of NMR data. The L^2 norm ($= \sqrt{(f_i^{ref} - f_i^{rec})^2}$) in this case describes the deviation between the reference spectrum (a singlet with no line-width) and the FM reconstructed equivalent, when selecting 256 of 1,024 complex data points in the time domain based on particular sampling strategy and seed value. One hundred seed values were used to create 100 specific sampling schedules for each sampling strategy. The L^2 values were hence sorted and the "best-of-100" (represented *rightmost* in the figure) and "worst-of-100" (represented *leftmost*) sampling schedules were identified for each strategy for further evaluation. (Note: for practical use, only the "best-of-100" schedule would be of interest)

2. Exponentially weighted random sampling

A modification to the uniformly random sampling is the exponentially weighted random sampling scheme. The rationale for this is that, by decaying signals, the signal intensity is higher at the beginning of the FID, yielding better sensitivity. The probability function is altered in an exponential fashion, so that it picks more points for shorter rather than longer evolution times. Randomization is used similarly to the uniformly random sampling. It should be mentioned that those samplings created by exponential weighting are a subset of the complete uniformly random one – the method only yields different results when a restricted subset is taken, such as a subset of 100 schedules as described above. Figure 1 shows that the exponentially weighted sampling schedule yields L^2 differences between 74 (worst) and 28 (best). Thus, this schedule is less likely to make big mistakes and has a greater probability of creating high-fidelity reconstructions. However, the variation of fidelity is still very large.

3. 5*Exponential sampling

Whereas the exponential sampling strategy matches the decay of the signal, attempts at optimizing the weighting of the probability function has been suggested. As it is common to apodize the reconstructed FID, the actual decay is faster than given by a single T_2. "Over-weighting" the probability function up to five times has hence come into question.

To improve schedules we considered the results from these various sampling strategies and tried to elucidate what makes a generally "good" sampling schedule, i.e., those with fewer aberrations than "bad" sampling schedules. We observed that when there are larger and/or more gaps in the beginning of the sampling schedule, the resulting reconstructions yield "dips" or "trenches" around the reconstructed peak. This is especially a problem when working with spectra with high dynamic range. These are naturally more pronounced for the uniformly random schedules than for the exponentially weighted ones. By "over-weighting" the exponential probability the problem is less severe. This would initially lead to the conclusion that "over-weighting" the sampling schedule strategy is better. However, just like apodization, it trades resolution for sensitivity – working contrary to the idea of NUS. Also, by "over-weighting" the probability function, the sampling scheme is biased towards sampling the very beginning of the FID. This raises the question as to whether it may be just as good simply to acquire the first part of the schedule uniformly, essentially falling back to uniform sampling (US) with fewer data points. Figure 1 demonstrates that the 5*exponential schedule exhibits the same worst-case scenario and disappointingly results in the worst overall performance.

4. US/NUS sampling

The considerations above have suggested that it may be good to sample the initial part uniformly (US), and then the rest of the schedule with NUS but in a uniformly-random point selection. As this introduces a discontinuity, many options are possible. For our search, we have typically sampled the first eighth of the schedule with US and the subsequent seven eighths with NUS and-uniformly random sampling point selection. Based on the point spread function analysis (Fig. 1) this schedule performs on average significantly better than the schedules discussed above. We find this strategy removes the artifacts, which manifest as "dips" or "trenches." However, to use this strategy one needs to make decisions as to how long to sample linearly and how to weight the NUS period. Thus, the selection appears to be arbitrary and is in general hard to optimize; however, further optimization of this approach may be worthwhile.

5. Poisson Gap sampling

As it is often desirable to use less or even much less than 50% sparse sampling where the number of non-obtained data coordinates are larger than the obtained data coordinates, it is valuable to look not only at the distribution of points to be acquired, but to study the nature and distribution of the gaps created in the sampling schedule. Doing so we found that (1) large gaps are more detrimental to the fidelity of the reconstruction and (2) gaps located in the beginning as well as in the end of the schedule impact the ability of good reconstruction. (Note, as it is most common to apply an apodization prior to transforming the time domain data to the frequency domain, the impact of gaps at the end of the schedule is much less than that at the beginning.) In addition, (3) the distribution should be sufficiently random in order to satisfy the Nyquist theorem on average.

When analyzing the distribution of gaps in large uniform sampling schedules we find that the gap sizes approach a Poisson distribution with the average gap size of 1/sparsity minus one. Here the sparsity is the fraction of recorded vs non-recorded data points. Thus, for a sparsity of 1/4, the average gap size would be $4 - 1 = 3$. Our experience is that it is typically easier to find a "good" sampling schedule with high fidelity reconstruction when using 25% sparse sampling of 8,000 data points rather than with 25% sparse sampling of 256 data points. We hypothesize that this is due to the stochastic nature of smaller sets. We further hypothesize that our ability to find a "good" sampling schedule increases when adding the constraint of a general Poisson distribution of the gaps in the sampling schedule. Generally, considering the observation that large gap sizes are more detrimental at the beginning and at the end of the sampling schedule, we can vary the local average of gap sizes during the sampling schedule so that the sampling is denser at the beginning and at the end of the schedule.. The selection of sampling points with Poisson gap sampling has recently been described in detail [34]. Figure 1 shows that this sampling schedule is least dependent on the seed number and has on average the lowest L^2 values.

7 Performance of Sampling Schedules on a Set of Four Peaks Using FM Reconstruction

So far, our analysis has been based on a single non-decaying signal without noise. This is certainly helpful to optimize the particular sampling schedule by entering different seed values into the underlying pseudo-random generator. However, it has little resemblance to the reality of NMR spectroscopy. In order to evaluate further the qualities of the above sampling strategies, we now simulate a set of four peaks that emulate a time domain acquisition to one T_2 (Figs. 2 and 3), and we extract subsets of the data with the particular sampling schedules. Next we reconstruct the time domain data with the FM procedure and create the 2D spectra with FFT. Finally we add synthetic Gaussian noise and evaluate reconstruction performance. The synthetic spectrum consists of one intense and well-separated peak, and three weak peaks, two of them almost overlapping.

Figure 2 shows FM reconstructions using the five NUS schedules discussed above and compares the spectra with the reference (top). The bottom spectrum is a simulation of traditionally US acquired data for a 1,024 complex data point FID, processed with cosine apodization, one set of zero filling and Fourier

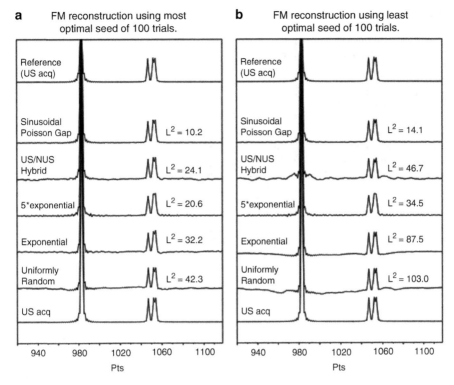

Fig. 2 Justification of prior selecting seed values and evaluation of sampling strategy based on a simulated spectrum with four signals, each simulated to been acquired to the equivalent of one T_2 and with no noise. (i.e., the simulated FID decays to 1/e.) The intensity of the *leftmost* signal is tenfold compared to each of the three others. (**a**) FM reconstruction using the sampling schedule based on labeled sampling schedules and seed value found to be "best-of-100" from previous evaluation (Fig. 1). (**b**) FM reconstruction using the sampling schedule based on labeled sampling schedules and using the seed value found to be "worst-of-100" from previous evaluation. The experience is described in the text

transformation. The leftmost signal has ten times the signal intensity of the three signals on the right. This is for evaluation of strong peaks, such as diagonal peaks in NOE spectroscopy (NOESY), and wherever a high dynamic range is exhibited. We chose a factor of 10 only for visibility purpose; we are aware that in real spectroscopy often a factor of 1,000:1 appears. The two rightmost signals provide a doublet, barely visible using uniform sampling. This arrangement allows for easy inspection of reconstruction fidelity and ability to preserve resolution.

In the left panel we use the sampling schedule with the best seed number as found in the data shown in Fig. 1. At the right side we use the seed numbers of the least favorable reconstruction. Thus, the two panels span the range of reconstruction fidelity consistent with the analysis of Fig. 1. The performance of the sampling schedules is most different for the least favorable seed numbers.

Fig. 3 Demonstration of initial state before FM reconstruction setting all non-obtained data values to zero, effectively using a discrete Fourier transformation (DFT) with the same sampling schedules as in Fig. 2a. The residual between reconstructed and reference here describe the combined point spread function from the four signals

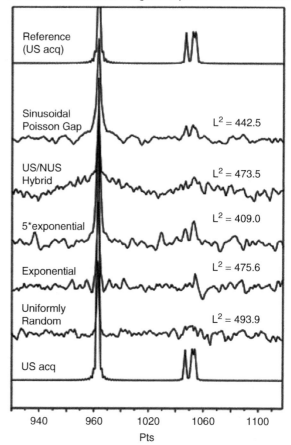

The second trace from the bottom (*Uniformly Random*) visualizes the same data as above; however, 256 of the 1,024 complex data points in the FID were chosen in a uniformly random fashion and FM reconstructed prior to apodization and further processing. Clearly in both the "best" and the "worst" cases, baseline abnormalities occur. These may not be detrimental in the case of spectra with more uniform size of signals, where these abnormalities are easily hidden in the noise, but are cause for concern for spectroscopy when the signals exhibit a high dynamic range. The third sections in both panels (*Exponential*) visualize the situation using exponential decaying sampling instead of uniformly random sampling. The baseline abnormalities are less pronounced, yet still present. Over-weighting the sampling using five times the exponent (*5*Exponential*) provides a flat but somewhat fluctuating baseline. However, the reduction of the baseline abnormalities has come at the price of lower resolution. The result of the *US/NUS hybrid* method is provided in the fifth section. The strategy does indeed provide a better baseline in the "best" case, with no apparent loss of resolution, yet is obviously in need of optimization of the particular sampling/seed; this fact is made evident by comparing the "best" case (Fig. 2) with the "worst" case. Here the *sinusoidal Poisson gap sampling* exhibits the best results (trace 6 from the bottom). Only deeper inspection can find differences between the reconstructed and the original data. Even the displayed "worst" reconstruction looks better than any of the other "best" cases. The top traces repeat the bottom traces (*US acq*) for visual reference.

8 Fourier Transformation of NUS Data Without Reconstruction of Missing Points

As it has been discussed in the literature, we also compare the performance of straight Fourier transformation FFT on the NUS synthetic four-line spectrum in Fig. 3. We used the schedules obtained with the best seed numbers as in Fig. 2a and leave the missing time domain data points at zero value. The traces presented in Fig. 3 show that the artifacts due to the sampling schedules and lack of reconstruction are severe and mask the small peaks. However, the intense line is readily observable. Thus, straight Fourier transformation may be an option if one is only concerned with very intense peaks, such as methyl resonances in a protein, and weak peaks are of little concern. This treatment of NUS data may be useful for a quick inspection of an NUS data set to find out whether an experiment has worked. However, it should be followed by a reconstruction effort to retrieve best the full information content of the NUS data.

9 Evaluation with Noise, Good and Bad

Experimental NMR spectra always contain noise. Hence, to evaluate the more realistic situation of the mentioned sampling schedules we add noise to the above test spectrum of four signals. The noise is generated with the NMRPipe utility *addNoise*; we tune the rms-value of the noise to a realistic situation generating a set of 100 different Gaussian distributions with seed values ranging from 1 through 100. To the reference situation of the traditionally uniformly sampled FID, we also add noise, but with twofold higher rmsd. This is to simulate the situation where equal time would be spent for data acquisition with uniform sampling (US) and non-uniform sampling (NUS). Since four times more scans can be accumulated per increment when only one quarter of the increments are measured, the noise-to-signal ratio in the time domain is twofold higher in the uniformly sampled spectrum. It is worth mentioning that even though the noise added in the time domain is of Gaussian or white noise character, the resulting distribution of the noise in the frequency spectrum is not Gaussian. This is due to apodization as well as non-uniform distribution of the noise in the NUS test cases.

We generated 100 noise test cases for each sampling strategy. This is because the effect of noise is intrinsically non-predictive. Reconstruction will work well in one distribution of the noise and less in another. Also, even though the initial noise is the same for each test case and seed value, different points of the noise will be sampled when the sampling schedule is applied. Hence, for a fairer comparison it is important not to draw a conclusion using just one set of noise.

In order to distinguish better and worse reconstructions, again an L^2 norm is applied. If the norm were applied to the whole spectrum, only noise itself would be evaluated which would unfairly favor the NUS cases. Hence, we applied the L^2 norm evaluation only to a very restricted area around the three small signals of interest shown bracketed with two vertical bars in the top trace of Fig. 4a. The best-of-100, mean, and worst-of-100 reconstruction results based on this L^2 norm application are provided in Fig. 4a–c, respectively, and the calculated L^2 norms are given in the spectra.

Figure 4a provides the considered best-of-100 results. The signals can easily be detected in each of the cases, and the splitting between the two adjacent signals can be seen to various degrees in all of the situations. The values of L^2 improve from the bottom to the top. Figure 4b illustrates the case for average (mean) noise. The splitting between the two overlapping signals has mostly vanished, or cannot be fully confirmed based on surrounding noise. The more isolated signal of the three small peaks is present in all traces, except possibly in the exponential weighted case. The values of the L^2 norm vary more than for the best case. Here, the 5*exponential distribution performs best. Figure 4c visualizes the situation when the reconstruction picks the least favorable noise set. For each of the sampling cases, the three small signals are hard to identify and barely resemble the spectrum without noise. The L^2 values are hence greater than in the mean- and the best-case scenarios. Note, the rightmost peak in the US spectrum, worst-case scenario, appears to have a shift in its position. However, the strongest peak still lines up with the reference. Shift of signal positions is something that has been associated with NUS, but evidently occurs in uniformly sampled spectra as well when the S/N ratio (SNR) is low. Of the reconstructed NUS spectra, it seems that the US/NUS sampling schedule features a similar shift of the position; the situation is however not worse in the NUS reconstructed spectra than in the case of traditional uniform sampling. Of interest is that the traditional US acquisition (bottom traces) has a tendency to produce false positive peaks. Note that this analysis applies to the case when sampling is tested for equal measuring times for US and NUS. If NUS is used to reduce the total experimental time the difficulties with maintaining signal positions are proportionally greater.

The L^2 analysis is simplistic but provides an illustrative picture of the situation. A total of 42 frequency data points were used, reflecting about an equal number of points representing signals in the reference spectrum and adjacent noise. There is, however, a strong indication that the likeness to the reference spectrum is greater in the FM reconstructed NUS spectra than in a traditional US acquired spectrum. The spread between the best-case scenario (where the noise "cooperates") and the worst-case scenario is greater than the differences in sampling strategy. Note that

Applications of Non-Uniform Sampling and Processing

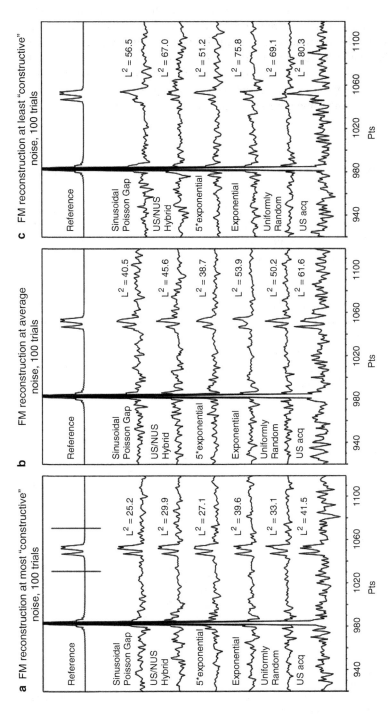

Fig. 4 FM reconstruction of the test situation in Fig. 2a – with noise added. The Gaussian noise was generated 100 times, each time with a different seed value and added in the time domain of the reference. The data was then sampled according to the various sampling schedules and FM reconstruction applied. In order to compare the situation of uniformly sampled data, the same noise was used, but with fourfold rmsd to emulate equal time acquisition. [Marked with *bars* in the evaluation is concentrated on the effect of the weaker peaks, only the data points 103 through 1,072 were used for L^2 norm evaluation. [Marked with *bars* in the reference spectrum in (**a**)]. As no prior knowledge about the noise, the result of the spectral presentation may vary from (**a**) when the noise is most cooperative, through (**b**) the average situation (median), to (**c**) where the noise is least cooperative

the particular sampling was optimized previously according to the method of using a single signal and no noise. There is, however, a trend that the sinusoidal Poisson gap sampling and the five times over-weighted exponential decay method are more faithful to the reference than the three other sampling strategies. The sinusoidal Poisson gap sampling strategy may offer better resolution, while the five times over-weighted strategy may produce somewhat better signal strength. Surprisingly, exponentially decaying sample weighting performs poorest in this evaluation.

10 Implementing NUS Schedules in Two Indirect Dimensions

Extending the probability weighted sampling strategies to two indirect dimensions is commonly done by creating a matrix and simply multiplying the weights from the two axes for probabilities of the individual points. The matrix is then converted into a one-dimensional object by taking the rows and concatenating them together. This is appropriate for all but the Poisson gap sampling strategy.

As Poisson gap sampling does not associate probabilities to a point but defines a relation between points in a one-dimensional string, a different approach is required. We use individual strings "woven" together, as outlined in Fig. 5, which allow us to cover two dimensions. First, a schedule with a desired weighting is implemented in one column followed by two rows and again two columns and so on (Fig. 5 left). What the figure does not reveal is that the strands have to be "pulled back" when a beginning is not occupied as a gap start is otherwise not defined. This

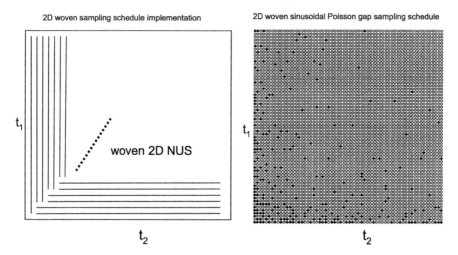

Fig. 5 Woven implementation of NUS Poisson gap sampling schedules in two indirect dimensions. First schedules are created with the selected schedule in the first t_1 column. This is followed by using the first two rows where the time points have not yet been selected. Subsequently, the next two columns are picked as indicated and so on

approach truly "weaves" the strands together. An example created with the "woven" Poisson gap sampling strategy is shown in Fig. 5 on the right.

11 Comparison of Sampling Schedules in Two Dimensions as Used in 3D NMR Spectra

In Fig. 6 we compare different sampling schemes for a matrix of nine synthetic peaks. Figure 6a shows the 2D spectra without noise. The left panel shows the entire 2D spectrum; the subspectrum that contains the peaks and was analyzed with the L^2 norm is indicated with a box. The subspectrum consists of a 31 × 31 point section of the middle part of the total 128 × 128 frequency point grid where the 2D frequency domain is determined from a 64 × 64 (or 4k) hyper complex matrix in the time domain. The signals of the matrix have increasing relative intensities, starting at a value of 2 and ending with an intensity of 10. The order of intensities is written next to the peaks in the rightmost panel. The initial synthetic time domain data set was created with the signals exponentially decaying to a value at point 64 of 1/e of its initial value in both dimensions to create 2D Lorenzian line shapes.

Cosine apodization and zero filling to 128 points were applied to the time domain data prior to processing in both dimensions. A common exponential stack is used for plotting contour levels, and the multiplier from one level to the next is 1.5. The two panels in the middle and at the right are plotted with different lowest starting levels to visualize the different intensities. The plot denoted with *6.0 is plotted on six times higher minimal level than the right hand plot. Thus, four more levels should be visible for each peak in the rightmost plot relative to the center one.

To simulate an appearance of real spectra, 100 sets of Gaussian noise were created, each in a 64 by 64 hypercomplex matrix and added to the synthetic time domain data. In Fig. 6b, d, uniform sampling (top left) is compared with different non-uniform sampling schedules. We are interested in comparing the situation where equal measuring time is used for US and NUS. Thus, when only measuring 1/16th of the increments, 16 times more scans can be acquired for each increment. For a fair comparison, and since noise adds non-coherently, we have added the same set of noise four times to the data set used to demonstrate US with FFT processing. Alternatively, we could have added 16 different sets of noise. Only one set of noise was added to the data sets used for simulating NUS and FM reconstruction. Five NUS strategies were evaluated, and in each case only 6.25% (1/16th) of the grid points were selected according to the sampling strategies. In other words, 256 of 4,096 hypercomplex time domain sampling points were used, selected based on sampling schedules optimized with the respective strategy and L^2 norm point spread function of a singlet without noise.

Each NUS data set selected for the seed number that gave the smallest L^2 value was combined with each of the 100 noise spectra and then FM reconstructed. The data were then processed with cosine apodization in both dimensions, including the

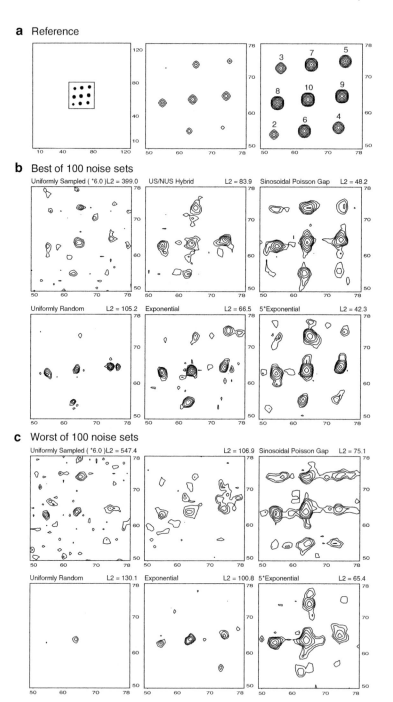

Fig. 6 Comparison of sampling schedules on a simulated spectrum of two indirect dimensions including Gaussian noise. Nine peaks of relative intensities 2–10 were created. (**a**) Complete spectrum simulated. The central area containing the peaks is boxed and expanded in the panels of

uniformly sampled data, and an L^2 evaluation was made in the 31 by 31 point subarea displayed in Fig. 6. The "best-of-100" (spectra with the smallest L^2 deviation from the reference) are shown in Fig. 6b. The "worst-of-100" are shown in Fig. 6c. The contours are plotted at the 1.0 level as defined in Fig. 6a except for the US panels (top left), where the noise was so overwhelming that the data had to be plotted at a minimum level six times the typical level for the NUS (*6.0).

The simulations presented in Fig. 6 show that, with two indirect dimensions, NUS has quite significant benefits in terms of signal to noise and the ability to detect weak peaks. Among the sampling schedules, 5*Exponential and sinusoidal Poisson gap sampling outperform the other sampling schedules. At least seven, and potentially all nine, of the peaks can be observed in the above two cases, while US detects only four or five at best. This is consistent with a previous report that NUS can enhance sensitivity [34].

12 Comparison of NUS in a 3D NOESY Spectrum

To evaluate sampling schedules on a 3D ^{15}N dispersed NOESY spectrum of a protein we recorded high-resolution data sets on ^{15}N labeled translation initiation factor eIF4E (Fig. 7). As a reference, the data was transformed with standard cosine apodization, zero filling, and FFT. A representative slice and cross section along the indirect ^{1}H dimension are shown in Fig. 7a. Reducing the time domain data to one third and one tenth with uniform sampling obviously leads to low resolution in the indirect dimension as shown in Fig. 7b1, b2. This is compared with NUS spectra following exponential weighting (Fig. 7c) and sinusoidal Poisson gap sampling (Fig. 7d). The 2D sinusoidal Poisson gap sampling schedule was created in a "woven" fashion as indicated in Fig. 5.

The comparison of the NUS spectra with the reference shows that reduction of the number of acquired increments to 30% results in good quality spectra, comparable to the reference and reduction of measurement time to one third. This is highly significant considering the fact that typical US 3D NOESY spectra for proteins are

Fig. 6 (continued) the rest of the figure. Contour plots of the central part plotted at two different starting levels of lowest contours. The contours (middle and right panel) are spaced by a factor of 1.5, and the relative intensity is indicated in the right-hand panel. (**b, c**) Peak recovery with the best (**c**) and worst (**d**) of the 100 Gaussian noise sets. As best and worst noise sets we selected those that resulted in the lowest and highest L^2 values for the selected area containing the peaks. This indicates the quality range of peak recovery one can expect. The six panels in (**c**) and (**d**) correspond to the same total number of scans. Thus, 16 times more scans per increment can be collected for the NUS data. To take account of this we multiplied the added noise by a factor of four for the US spectra (*top left panel*). The simulations show that NUS recovers the peaks in the noisy spectrum significantly better than US when equal measuring times are considered. For all NUS schedules the best of 100 seed numbers were used as selected with minimizing the point spread function (**c**)

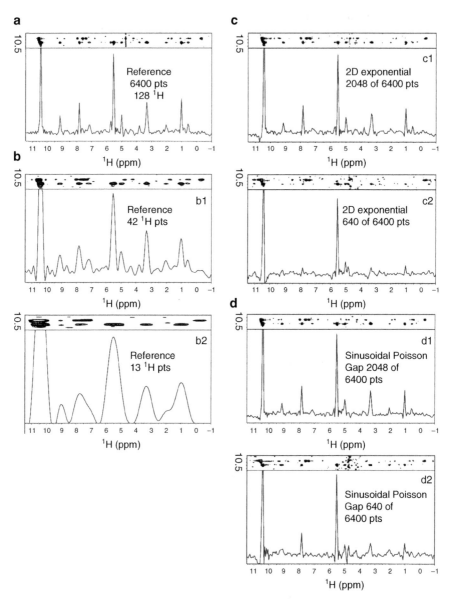

Fig. 7 Comparison of alternative processing on a 3D ^{15}N-NOESY-HSQC spectrum of human translation initiation factor eIF4e. (**a**) Uniformly sampled reference. The time domain data were acquired as 6,400 hyper-complex points sampled in the two indirect dimensions [128 (H$_{indir}$) × 50 (^{15}N)]. The spectra were measured on a 700-MHz spectrometer with sweep widths of 9765 Hz and 2270 Hz, respectively. The t_{max} hence were 0.013 and 0.022 s each for the indirect proton and nitrogen dimensions, respectively, representing nearly an optimal situation for the nitrogen dimension, but not for the indirect proton dimension. Data were transformed with the standard FFT procedure after cosine apodization and doubling the time domain by zero filling. (**b**) Reducing the number of complex points to 42 (32%) (**b1**) and 13 (10%) (**b2**) in the indirect proton dimension, cosine apodization, and zero filling result in low resolution spectra in the indirect

recorded for several days. On the other hand, reduction of sampling points to 10% (Fig. 7c2, d2) leads to loss of weak peaks. The NUS schedules of Fig. 7 use equal dilution of sampling in both indirect dimensions. It seems that less dilution in the proton dimension and a more drastic dilution in the heteronuclear (^{15}N) dimension should be explored and might lead to superior results.

13 Discussion

Uniform sampling of multidimensional NMR spectra requires a large amount of measuring time if the resolution promised by high field spectrometers is to be fully utilized. Paradoxically, if one obtains a higher field magnet and maintains the same number of increments as on a lower field spectrometer, resolution is lost because one does not reach the same maximum evolution time as the dwell time is shortened. Thus the spectroscopist has to record more increments to utilize the increased resolving power of the higher field instrument. Obviously this requires more measuring time. Ideally one wants to sample up to $1.2*T_2$ of the respective coherence in each indirect dimension [8]. This is an unacceptable situation, and replacing US with more economic alternatives has been a widely accepted goal of NMR development.

The first NUS approaches have used experimentally weighted random sampling methods and variations of random sampling. However, other NUS methods, such as radial sampling with projection reconstruction, have been proposed (see introduction). Different algorithms have been used for reconstruction of NUS data, such as Maximum Entropy or Maximum Likelihood Methods. It is beyond the scope of this chapter to compare exhaustively the different approaches. Here we use just the FM reconstruction software to compare the performance of different random sampling schedules where the randomness is skewed by weighting functions.

We have shown previously that the resolution of triple resonance experiments can be dramatically enhanced with random NUS in the indirect dimensions, and high-resolution 3D or 4D triple-resonance spectra of large proteins can be recorded within a few days, which would otherwise require months of instrument time with US [32, 35]. This works rather well with triple resonance experiments where peaks have similar intensities and there is not much of a sensitivity and dynamic range issue. In fact, if there are primarily strong signals, such as methyl peaks in ILV labeled samples, a straight Fourier transformation of the NUS data where the

---

Fig. 7 (continued) dimensions. (**c**) FM reconstruction of 2,048 (32%) (**c1**) and 640 (10%) (**c2**) data points sampled with an exponential weighting schedule in the two indirect dimensions. (**d**) FM reconstruction of 2,048 (32%) (**d1**) and 640 (10%) (**d2**) sampled data points according to weaved sinusoidal Poisson gap sampling. For all spectra, equal numbers of scans were recorded per increment. Thus, the NUS spectra were acquired in one third and one tenth of the time used for the US spectrum, respectively

missing data points are left at zero may faithfully retrieve the positions of the strongest peaks (see Fig. 3) but weaker signals are lost.

The focus of this chapter is to evaluate different sampling schedules on samples that have a sensitivity issue, and we have evaluated sampling schedules assuming equal total measuring times. We investigated five sampling strategies: uniformly random, exponentially decaying random sampling, fivefold steeper decaying exponential weighting, a hybrid of US and NUS, and sinusoidal Poisson gap sampling. We find that the performance of any sampling schedule created with the help of a random number generator depends crucially on the original seed number. To determine optimal seed numbers we use the point-spread function, which determines the quality of the reconstruction by calculating the L^2 norm, the squared difference between a synthetic signal and the reconstruction using a particular sampling schedule.

Among all sampling schedules tested, we find that sinusoidal Poisson gap sampling depends least on the choice of the seed number but for the best seed numbers the different sampling schedules produce nearly identical results (Fig. 1). The different sampling schedules were then compared for situations without and with noise. Random noise has also to be created with seed numbers, and the noise can interfere with signals constructively (good noise) or destructively (bad noise). Thus, using both situations we have explored the range of reconstruction performances (Figs. 4 and 6).

To implement NUS schedules for Poisson gap sampling in two indirect dimensions we use a "woven" selection by alternate picking of numbers in two columns followed by two rows and two columns and so on. Sampling times in each row or column are picked according to a desired sampling schedule. In this way we can evaluate sampling performance in two indirect dimensions. A similar weave NUS scheduling can also be used in higher dimensions.

Most of the evaluation of sampling schedules was done with simulated spectra and noise assuming equal total instrument time. This indicates that when using the time gained by not sampling all Nyquist grid points, and using the time gained for measuring more transients for the sampled points, one can obtain a better signal-to-noise ratio and detect peaks that are otherwise lost in the noise, which corresponds to a sensitivity gain (Fig. 6). This is consistent with a previous observation with ^{13}C detected spectra [34].

Finally, we have compared different sampling schedules on an experimental ^{15}N dispersed NOESY of a protein, where an equal numbers of scans were recorded per increment, and using NUS results in a shorter measuring time (Fig. 7). Here it is obvious that, with one third of the increments, one can recover nearly the same quality spectrum as obtained with the threefold longer US acquisition. It seems, however, that in heteronuclear-dispersed NOESY spectra it is worthwhile to maintain a higher sampling density in the ^{1}H dimension than in the heteronuclear dimension.

Acknowledgment This research was supported by the National Institutes of Health (grants GM 47467 and EB 002026). We thank Bruker Biospin for providing access to a 700-MHz spectrometer for acquiring the experimental data.

References

1. Ernst RR, Anderson WA (1966) Application of Fourier transform spectroscopy to magnetic resonance. Rev Sci Instr 37:93–106
2. Cooley JW, Tukey JW (1965) An algorithm for the machine calculation of complex Fourier series. Math Comput 19:297–301
3. Barna JCJ, Laue ED, Mayger MR, Skilling J, Worrall SJP (1987) Exponential sampling, an alternative method for sampling in two-dimensional NMR experiments. J Magn Reson 73:69–77
4. Sibisi S, Skilling J, Brereton RG, Laue ED, Staunton J (1984) Maximum entropy signal processing in practical NMR spectroscopy. Nature 311:446–447
5. Hoch JC (1989) Modern spectrum analysis in nuclear magnetic resonance: alternatives to the Fourier transform. Meth Enzymol 176:216–241
6. Frueh DP, Sun ZY, Vosburg DA, Walsh CT, Hoch JC, Wagner G (2006) Non-uniformly sampled double-TROSY hNcaNH experiments for NMR sequential assignments of large proteins. J Am Chem Soc 128:5757–5763
7. Rovnyak D, Frueh DP, Sastry M, Sun ZY, Stern AS, Hoch JC, Wagner G (2004) Accelerated acquisition of high resolution triple-resonance spectra using non-uniform sampling and maximum entropy reconstruction. J Magn Reson 170:15–21
8. Rovnyak D, Hoch JC, Stern AS, Wagner G (2004) Resolution and sensitivity of high field nuclear magnetic resonance spectroscopy. J Biomol NMR 30:1–10
9. Schmieder P, Stern AS, Wagner G, Hoch JC (1993) Application of nonlinear sampling schemes to COSY-type spectra. J Biomol NMR 3:569–576
10. Schmieder P, Stern AS, Wagner G, Hoch JC (1994) Improved resolution in triple-resonance spectra by nonlinear sampling in the constant-time domain. J Biomol NMR 4:483–490
11. Schmieder P, Stern AS, Wagner G, Hoch JC (1997) Quantification of maximum-entropy spectrum reconstructions. J Magn Reson 125:332–339
12. Shimba N, Stern AS, Craik CS, Hoch JC, Dotsch V (2003) Elimination of ^{13}Calpha splitting in protein NMR spectra by deconvolution with maximum entropy reconstruction. J Am Chem Soc 125:2382–2383
13. Sun ZJ, Hyberts SG, Rovnyak D, Park S, Stern AS, Hoch JC, Wagner G (2005) High-resolution aliphatic side-chain assignments in 3D HCcoNH experiments with joint H-C evolution and non-uniform sampling. J Biomol NMR 32:55–60
14. Sun ZY, Frueh DP, Selenko P, Hoch JC, Wagner G (2005) Fast assignment of ^{15}N-HSQC peaks using high-resolution 3D HNcocaNH experiments with non-uniform sampling. J Biomol NMR 33:43–50
15. Tugarinov V, Kay LE, Ibraghimov I, Orekhov VY (2005) High-resolution four-dimensional 1H-13C NOE spectroscopy using methyl-TROSY, sparse data acquisition, and multidimensional decomposition. J Am Chem Soc 127:2767–2775
16. Chylla RA, Markley JL (1995) Theory and application of the maximum likelihood principle to NMR parameter estimation of multidimensional NMR data. J Biomol NMR 5:245–258
17. Marion D (2005) Fast acquisition of NMR spectra using Fourier transform of non-equispaced data. J Biomol NMR 32:141–150
18. Gutmanas A, Jarvoll P, Orekhov VY, Billeter M (2002) Three-way decomposition of a complete 3D 15N-NOESY-HSQC. J Biomol NMR 24:191–201
19. Hiller S, Ibraghimov I, Wagner G, Orekhov VY (2009) Coupled decomposition of four-dimensional NOESY spectra. J Am Chem Soc 131:12970–12978

20. Korzhneva DM, Ibraghimov IV, Billeter M, Orekhov VY (2001) MUNIN: application of three-way decomposition to the analysis of heteronuclear NMR relaxation data. J Biomol NMR 21:263–268
21. Orekhov VY, Ibraghimov I, Billeter M (2003) Optimizing resolution in multidimensional NMR by three-way decomposition. J Biomol NMR 27:165–173
22. Orekhov VY, Ibraghimov IV, Billeter M (2001) MUNIN: a new approach to multi-dimensional NMR spectra interpretation. J Biomol NMR 20:49–60
23. Coggins BE, Venters RA, Zhou P (2005) Filtered backprojection for the reconstruction of a high-resolution (4,2)D CH3-NH NOESY spectrum on a 29 kDa protein. J Am Chem Soc 127:11562–11563
24. Coggins BE, Zhou P (2006) Polar Fourier transforms of radially sampled NMR data. J Magn Reson 182:84–95
25. Coggins BE, Zhou P (2008) High resolution 4-D spectroscopy with sparse concentric shell sampling and FFT-CLEAN. J Biomol NMR 42:225–239
26. Kazimierczuk K, Kozminski W, Zhukov I (2006) Two-dimensional Fourier transform of arbitrarily sampled NMR data sets. J Magn Reson 179:323–328
27. Kazimierczuk K, Zawadzka A, Kozminski W, Zhukov I (2006) Random sampling of evolution time space and Fourier transform processing. J Biomol NMR 36:157–168
28. Kim S, Szyperski T (2003) GFT NMR, a new approach to rapidly obtain precise high-dimensional NMR spectral information. J Am Chem Soc 125:1385–1393
29. Kupce E, Freeman R (2004) Fast reconstruction of four-dimensional NMR spectra from plane projections. J Biomol NMR 28:391–395
30. Kupce E, Freeman R (2004) Projection-reconstruction technique for speeding up multidimensional NMR spectroscopy. J Am Chem Soc 126:6429–6440
31. Venters RA, Coggins BE, Kojetin D, Cavanagh J, Zhou P (2005) (4,2)D Projection–reconstruction experiments for protein backbone assignment: application to human carbonic anhydrase II and calbindin D(28K). J Am Chem Soc 127:8785–8795
32. Hyberts SG, Frueh DP, Arthanari H, Wagner G (2009) FM reconstruction of non-uniformly sampled protein NMR data at higher dimensions and optimization by distillation. J Biomol NMR 45:283–294
33. Hyberts SG, Heffron GJ, Tarragona NG, Solanky K, Edmonds KA, Luithardt H, Fejzo J, Chorev M, Aktas H, Colson K et al (2007) Ultrahigh-resolution (1)H-(13)C HSQC spectra of metabolite mixtures using nonlinear sampling and forward maximum entropy reconstruction. J Am Chem Soc 129:5108–5116
34. Hyberts SG, Takeuchi K, Wagner G (2010) Poisson-gap sampling and forward maximum entropy reconstruction for enhancing the resolution and sensitivity of protein NMR data. J Am Chem Soc 132:2145–2147
35. Frueh DP, Arthanari H, Koglin A, Walsh CT, Wagner G (2009) A double TROSY hNCAnH experiment for efficient assignment of large and challenging proteins. J Am Chem Soc 131:12880–12881

Erratum to: Data Sampling in Multidimensional NMR: Fundamentals and Strategies

Mark W. Maciejewski, Mehdi Mobli, Adam D. Schuyler, Alan S. Stern, and Jeffrey C. Hoch

Erratum to: Top Curr Chem
 DOI: 10.1007/128_2011_185

The author list for references 44 and 45 has not been published correctly.

The correct author list for ref. 44 is Mobli M, Stern AS, Hoch JC
The correct author list for ref. 45 is Mobli M, Stern AS, Bermel W, King GF, Hoch JC

10.1007/128_2011_291

The online version of the original chapter can be found under
DOI: 10.1007/128_2011_185

M.W. Maciejewski, A.D. Schuyler, and J.C. Hoch (✉)
Department of Molecular, Microbial, and Structural Biology, University of Connecticut Health Center, 263 Farmington Ave, Farmington, CT 06030-3305, USA
e-mail: hoch@uchc.edu

M. Mobli
Division of Chemistry and Structural Biology, Institute for Molecular Bioscience, The University of Queensland, St. Lucia, Brisbane, QLD 4072, Australia

A.S. Stern
Rowland Institute at Harvard, 100 Edwin H. Land Blvd., Cambridge, MA 02142, USA

Index

A
Accordion, 59
Additive algorithm, 7
Algebraic algorithm, 10
Aliasing, 62
APSY, 21, 26

B
Bandwidth, 62
Bayesian inference, 1, 14
Bracewell's Fourier transform slice/projection theorem, 1
Burst sampling, 72

C
CLEAN algorithm, 101
Correlation peaks, 6
Covariance spectroscopy, 82
Cross peaks, projections, 25
Cube MFT, 112

D
Dark spectrum, 101
Discrete Fourier transform (DFT), 50, 136
Dynamic nuclear polarization (DNP), 81

E
Eclipsed resonances, 12
Evolution space, 15

F
Fast Fourier transform (FFT), 128, 130
Filtered back-projection, 8
FM reconstruction, 125
Fourier pair, 83
Fourier transform, 83
  multidimensional, 85
Free induction decay (FID), 50

G
GAPRO, 21, 23
Gridding, 100

H
Hybrid schemes, 10
Hypercomplex Fourier transform, 5
Hyperdimensional spectroscopy, 1

I
Integration, 97
Inverse Fourier transform (IFT), 83

L
Linearly increasing concentric ring sampling (LCRS), 96
Loop gain, 103
Lowest amplitude, 9
Lowest-value algorithm, 9

M
Markov chain Monte-Carlo, 14
Maximum entropy (MaxEnt) reconstruction, 55, 100, 127
Maximum likelihood, 55, 126
MFT, sparse, 110

Monte Carlo methods, 14
Multidimensional decomposition (MDD), 55, 126

N
NMR, 1ff
    biomolecular, 79
    hyperdimensional, 1
    multidimensional, 1, 79
    three-dimensional, 3
NOESY, 125, 135
    3D, 143
    heteronuclear-edited, 118
Noise, 89, 104, 108, 137
Nonlinear sampling, 79
Nonuniform averaging, 72
Nonuniform sampling (NUS), 49, 62, 125, 137
NUS explosion, 61

O
Occultation, 2
Oversampling, 52

P
Parallax, 2
Peak filter, secondary, 29
Peak picking, automated, 21
Plane projections, 17
Point-spread function, 55, 131
Poisson gap sampling, 72, 133, 137
Probability, weighted, 97
Projected linewidths, 13
Projection angles, selection, 31
Projection–cross-section theorem, 23
Projection-reconstruction, 1
Projection spectroscopy, 21
Protein backbone, 21
Proteins, 115
    side chain resonances, 41
Protein side chains, 21

Q
Quadrature detection, 53

R
Random phase detection, 73
Random sampling, 59
Reconstruction, 6
    deterministic, 6
Redux, 60
Residual dipolar couplings (RDCs), 54
Resonance assignments, automated, 21
    APSY-based, 35
RNA polymerase, 115

S
Samples, weighted, 97
Sampling, artifacts, 62, 101
    beat-matched, 71
    concentric rings, 96
    noise, 108
    nonuniform (NUS), 49, 62, 125, 130, 137
    on-grid/off-grid, 62
    radial, 3, 61, 95
    random, 59, 96
    sparse, 1, 79, 98
    spiral, 70, 96
    uniform (US), 137
Signal detection, sensitivity, 29
Signal-to-noise (S/N) ratio, 3, 52, 59, 90, 97
Slice/projection theorem, 1, 4
Sparse sampling, 1, 79, 98
Spectroscopy, four-dimensional, 15
    hyperdimensional, 17
Spectrum analysis, 49
Spiral sampling, 70, 96

T
Time-proportional phase incrementation (TPPI), 53, 73
Transform kernel, 83
Triple resonance, 125

U
Ubiquitin, 116
Uniform sampling (US), 137

X
X-ray tomography, 2